Lecture Notes in Energy

Volume 35

Lecture Notes in Energy (LNE) is a series that reports on new developments in the study of energy: from science and engineering to the analysis of energy policy. The series' scope includes but is not limited to, renewable and green energy, nuclear, fossil fuels and carbon capture, energy systems, energy storage and harvesting, batteries and fuel cells, power systems, energy efficiency, energy in buildings, energy policy, as well as energy-related topics in economics, management and transportation. Books published in LNE are original and timely and bridge between advanced textbooks and the forefront of research. Readers of LNE include postgraduate students and non-specialist researchers wishing to gain an accessible introduction to a field of research as well as professionals and researchers with a need for an up-to-date reference book on a well-defined topic. The series publishes single and multi-authored volumes as well as advanced textbooks.

More information about this series at http://www.springer.com/series/8874

C.E. (Sandy) Thomas

Stopping Climate Change: the Case for Hydrogen and Coal

Springer

C.E. (Sandy) Thomas
Former President (ret) of H2Gen
Innovations, Inc.
Alexandria, VA
USA

ISSN 2195-1284 ISSN 2195-1292 (electronic)
Lecture Notes in Energy
ISBN 978-3-319-31654-3 ISBN 978-3-319-31655-0 (eBook)
DOI 10.1007/978-3-319-31655-0

Library of Congress Control Number: 2016950890

Printed on acid-free paper

This Springer imprint is published by Springer Nature
The registered company is Springer International Publishing AG
The registered company address is: Gewerbestrasse 11, 6330 Cham, Switzerland

Foreword

Many of us—engineers, economists, also environmentalists, the political class and a growing group of lay people—consider two challenges dominant in their thinking and acting: Climate change and the urgent reduction of world dependence on oil imports from politically and economically not too stable nations and their oligopols.

Humans' influence on climate change is the consequence of their releasing large amounts of carbon dioxide (CO_2), methane (CH_4) and other emissions to the earth's atmosphere along the complete energy conversion chain from the primary energy raw materials coal, oil, gas via primary and secondary energies ending up at a great variety of energy services such as heat, electricity, transport and information. Each chain link contributes emissions, some more, some less. The dominant emitters by far are coal fired power stations, the steel and cement industries. By contrast, intermittent renewable and nuclear energies are operational without contributing to climate change.

What we face is finding ways to provide the economically imperative amounts of environmentally clean energy services without raising their influence on climate change. In essence, modern efficient technologies are indispensable, they have to guarantee more clean and climate change irrelevant energy services with less polluting and climate changing primary energy raw materials! The service kilowatt-hour not asked for in households, on the road, in industry and trade because of efficiency increases along each step of the energy conversion chain avoids input of polluting and climate changing energy raw materials at the beginning of the chain.

Dr. C.E. (Sandy) Thomas' new book, entitled "Stopping Climate Change—the Case for Hydrogen and Coal," advocates his unmistakable credo: Cut back politically fragile, environmentally polluting and climate changing oil imports; put at the top of the world's agenda renewables wind, solar, biomass, hydro, and—the most important—don't burn coal! Instead, gasify it and generate power for the ever increasing electrification of humans' energy world using hydrogen in a series of consecutive HT-fuel cells, gas turbines, steam turbines, all together a convincingly efficient process: Integrated gasification combined cycle (IGCC). The co-produced

CO_2 is safely and securely sunk for final storage or later use in underground obsolete or emptied coal and oil fields or in saline aquifers.

Life Cycle Assessments (LCA) say that a coal mine can have an operational life of up to 100 years, and coal fired power stations somewhere near 40–50 years. Consequently, the bitterly needed decarbonization of the energy economy is nothing for a jump start! Decarbonization and hydrogenation (less carbon, more hydrogen) are priorities for generations working in development shops and industry. Hydrogen, electrolyzed with the help of renewable electricity, or made from biomass or coal and reelectrified in fuel cells, delivers a combination of clean and climate neutral electricity and heat. After manual energy of first humans on earth, after mechanical energy (the wheel), after early renewable technologies (sailboats, waterwheels, windmills, well insulated dwellings), after heat and steam engines, electric motors and gasoline engines, hydrogen energy and fuel cells are the next steps to be taken up humans' energy stairway.

We wish Sandy's book in the hands of energy engineers and economists and, of course, its content into their thinking and will. It is also a trustworthy educational resource for energy lay persons and the political class. Don't look for it in dusty bookshelves, because the book belongs on your desk—open for immediate consultation.

Carl-Jochen Winter
University of Stuttgart, and Former Executive Board Member
of the German Aerospace Center (DLR)

Acknowledgements

My interest in hydrogen-powered fuel cell electric vehicles was first stimulated by the visionary work of Prof. Joan Ogden[1] via her 1989 booklet with Robert Williams from Princeton for the World Resources Institute entitled "Solar Hydrogen: Moving Beyond Fossil Fuels". At the time I was a legislative assistant to Senator Tom Harkin of Iowa. Senator Harkin accelerated the development of hydrogen as an energy carrier when he succeeded in doubling the early hydrogen research budget at the Department of Energy…from \$1 million to \$2 million. The DOE's hydrogen budget subsequently rose to over \$200 million per year, but it was Senator Harkin's influence along with Representative Bob Walker of Pennsylvania that kept the Federal hydrogen energy research budget growing through Congressional appropriations in the early years of hydrogen development.

I would also like to thank: Dr. Bob Shaw, and the other investors in H_2Gen Innovations, Inc. of Alexandria, Virginia; Dr. Shaw was the lead investor in H_2Gen, the company we set up to develop and commercialize steam methane reformers to convert natural gas to hydrogen at the fueling station; Barney Rush, the CEO of H_2Gen also reviewed early chapters and made valuable suggestions related to electric utility systems; Ira Kuhn, Jr., then the President of Directed Technologies, Inc. (DTI) of Arlington, Virginia, Ira recognized the potential of hydrogen as an energy carrier early on, and convinced the Ford Motor Company in 1993 to undertake a joint development with DTI for the U.S. DOE to explore hydrogen-powered FCEVs and the means to safely deliver affordable hydrogen to those vehicles; Dr. Sigmund Gronich, who led the hydrogen and fuel cell development programs at the DOE for many years; Dr. Gronich first suggested to this author that coal would be a good source of hydrogen; Brian D. James who has been conducting in-depth mass production cost estimates of fuel cells and hydrogen

[1] In 1989 Dr. Ogden was Professor at Princeton University; she has since moved to the Institute for Transportation Studies at University of California at Davis where she continues her research on hydrogen and fuel cells, training a new batch of young students on the merits of hydrogen as an energy carrier in the process.

energy equipment for the Ford Motor Company and the Department of Energy when he worked at DTI, continuing today with Strategic Analysis, Inc.; Dr. Frank Lomax was the backbone of the scientific and engineering efforts at both DTI and H_2Gen to develop affordable hydrogen energy systems; Ron Sims & Brad Bates then at the Research Department of the Ford Motor company in Dearborn, Michigan, who helped set up and run the DOE/DTI/Ford hydrogen and FCEV Development program; Bob Rose, the leader of the Fuel Cell Council, an early industry association dedicated to the development of fuel cell systems and Jeff Serfass, the former President of the National Hydrogen Association (NHA)[2] set up to promote the commercial uses of hydrogen as an energy carrier; Amgad Elgowainy has provided detailed guidance on the cost of various hydrogen system components at the Argonne National Laboratory; Frank Novachek with Excel Energy, a large Midwestern utility, Frank led the group development of several explorations of the merits of hydrogen and fuel cells for the NHA, He also reviewed many of the chapters in this book and made many valuable suggestions for improvements in the text.

[2]The Fuel Cell Council and the NHA were subsequently merged in 2010 to form the Fuel Cell and Hydrogen Energy Association a trade association led by Morry Markowitz

Contents

Abbreviations

ADG	Anaerobic Digester Gas
AEO	Annual Energy Outlook (DOE publication)
AER	All Electric Range
AFV	Alternative Fuel Vehicle
ANL	Argonne National Laboratory
BEV	Battery Electric Vehicle
CARB	California Air Resources Board
CC	Combined Cycle
CCS	Carbon Capture and Sequestration (or Storage)
CF	Capacity Factor
CO_2	Carbon Dioxide
DFMA	Design for Manufacturing and Assembly
DOE	(U.S.) Department of Energy
DTI	Directed Technologies, Inc
E-85	Mix of 85% ethanol in gasoline
EIA	Energy Information Administration (part of DOE)
EPA	(U.S.) Environmental Protection Agency
EtOH	Ethanol
EU	European Union
EV	Electric Vehicle
FCEV	Fuel Cell Electric Vehicle
GH2	Gaseous Hydrogen
GHG	Greenhouse Gas
GREET	Greenhouse gases, Regulated Emissions and Energy use in Transportation
H_2	Hydrogen
H2A	Department of Energy Hydrogen Cost program
HEV	Hybrid Electric Vehicle
HHV	Higher Heating Value
ICE	Internal Combustion Engine

ICV	Internal Combustion (engine) Vehicle
IEA	International Energy Agency
IGCC	Integrated Gasification Combined Cycle (coal-powered power plant)
IGFC	Integrated Gasification Fuel Cell (coal-powered power plant)
IOU	Investor-owned utility
IPCC	Intergovernmental Panel on Climate Change
IRR	Internal rate of return
kW	kiloWatt (unit of power equal to 1,000 watts)
kWh	kiloWatt-hour (unit of energy equal to 1,000 watt-hours)
LDV	Light Duty Vehicle
LH2	Liquid Hydrogen
Li-Ion	Lithium Ion (battery)
MBTU	Million btu (one btu equals approximately 1,055 joules of energy)
MCFC	Molten Carbonate Fuel Cell
MHE	Material Handling Equipment such as fork lift trucks
MIT	Massachusetts Institute of Technology
MMT	Million metric tonnes
NETL	National Energy Technology Laboratory
NG	Natural Gas
NGCC	Natural Gas Combined Cycle (power plant)
NGV	Natural Gas Vehicle
NiMH	Nickel Metal Hydride (battery)
NREL	National Renewable Energy Laboratory
NT SUV	Non-Truck SUV
O&M	Operation and maintenance (costs)
OPEC	Organization of Petroleum Exporting Countries
PEM	Proton Exchange or Polymer Electrolyte Membrane (type of fuel cell)
PHEV	Plug-in hybrid electric vehicle
PSA	Pressure Swing Adsorption (gas clean-up device)
PV	Photovoltaic (solar cell producing electricity from the sun)
RMI	Rocky Mountain Institute
SAI	Strategic Analysis Inc
SEIA	Solar Energy Industry Association
SMR	Steam Methane Reformer
SOFC	Solid Oxide Fuel Cell
SUV	Sport Utility Vehicle
USABC	US Advanced Battery Consortium
VMT	Vehicle Miles Traveled
WWTP	Wastewater Treatment Plant
ZEV	Zero Emission Vehicle

Chapter 1
Introduction to "Stopping Climate Change: The Case for Coal and Hydrogen"

Abstract There are two grave threats to modern civilization: Climate Change, and Continued reliance on imported oil from the volatile Middle East. Both of these threats can be ameliorated by the development of hydrogen and particularly hydrogen made from coal as a major energy carrier to replace oil products such as gasoline and diesel fuel to power our motor vehicleswith electricity made by gasifying coal (rather than burning coal) with carbon capture and storage/sequestration (CCS). Most discussions of climate change emphasize extreme weather events, rising temperatures and rising sea levels. While these aspects of climate change may be very disruptive, the real catastrophic impact of climate change may be our inability to provide enough food for a growing world population due to the confluence of higher temperatures that reduce food production, coupled with increased global consumption of meat that requires more crop production than eating fruits, vegetables and grains as people aspire to the living standards in developed nations, and the production of biofuels from crops (Biofuels can also be produced from waste products such as municipal solid waste (MSW) without impacting crop production.) that limits farm land available for food production. Powering our light duty vehicles with hydrogen made from coal, and our electricity from a combination of nuclear energy, renewables and natural gas and coal gasification with carbon capture and storage/sequestration (In Carbon capture and sequestration, the carbon in the form of carbon dioxide (CO_2) is recovered at the power plant, compressed, and shipped by pipeline to an underground storage cavern where it is buried underground, thereby preventing the CO_2 from entering the atmosphere where it would have exacerbated the impact of climate change.) (CCS) would drastically reduce or even eliminate our dependence on foreign oil, (The U.S. petroleum production is sufficient to supply all non-transportation oil requirements.) and, if the rest of the world follows our lead, could eliminate the influence of OPEC and our dependence oil imported from the volatile Middle East, and this approach would allow us to meet the climate change goal of reducing greenhouse gas (GHG) emissions to 80 % below 1990 levels by 2050 while relying primarily on our huge inventory of domestic coal to supply both hydrogen and electricity. Many advocates of using hydrogen as a zero-emission energy carrier envision a future where hydrogen is produced by electrolyzing water using renewable energy. While this vision of a renewable hydrogen economy may be

C.E. Thomas, *Stopping Climate Change: the Case for Hydrogen and Coal*,
Lecture Notes in Energy 35, DOI 10.1007/978-3-319-31655-0_1

a long-term goal for society, it might be too costly or take too long to make the transition from our current predominantly coal- and natural gas-based electricity generation system and our nearly total dependence on oil products for transportation to an all-renewable future in the next few decades. This book therefore explores a more plausible transition strategy using coal, natural gas, renewable and nuclear power as the primary sources of energy over the next few decades as a practical pathway toward a renewable hydrogen future. Making hydrogen from renewables by electrolyzing water also raises the issue of *water resources*. It may be necessary to develop technologies to electrolyze sea water to make renewable hydrogen a sustainable long-term option with respect to water utilization.In the near term, hydrogen made from coal offers an alternative less dependent on our planet's limited fresh-water resources. While this book is focused primarily on reducing greenhouse gas (GHG) emissions from electricity power plants, we also explore the option of jump-starting a near zero-emission hydrogen-powered fuel cell electric vehicle (FCEV) transportation system by installing "Tri-Generation" fueling stations at warehouses, distribution centers and "big box stores." These "Tri-Gen" stations generate three energy carriers made from natural gas: hydrogen, electricity and heat; Chap. 7 demonstrates that such distributed "Tri-Gen" stations would provide a significant return on investment in Hawaii, California and the Eastern portion of the United States by displacing expensive electricity at the stores, thereby allowing the profitable introduction of hydrogen fueling stations to ramp up the introduction of zero emission FCEVs in the nation's vehicle fleet. We show that these Tri-Gen stations could be profitable today by providing hydrogen at many warehouses that have already converted from battery-operated fork lift trucks to fuel cell lift trucks powered by hydrogen.

1.1 Climate Change Threats to Society

Most articles on climate change list the dangers to society in terms of more extreme weather including rising temperatures, droughts, wildfires floods, etc., along with rising seas levels that could submerge many island nation/states, and could cost many billions of dollars to protect major metropolitan cities that are adjacent to oceans. In January, 2016, NASA and the National Oceanic and Atmospheric Administration (NOAA) based on separate data sources announced that 2015 was the hottest year on record, surpassing 2014 which was the previous hottest year [1]. They also stated that the 11 hottest years on record have all occurred in the last 20 years.

However, climate change poses much greater risks to society than simply extreme weather events, rising sea levels and rising temperatures as dire as those threats may be. As Bourne [2] documents in his book "The End of Plenty", climate change may have much more calamitous consequences than rising temperatures or rising sea levels for the world in terms of our ability to feed all inhabitants of the planet.

Bourne suggests three forces that together could severely hamper our ability to feed the citizens of Planet Earth:

- Population Growth
- Increased meat consumption
- Bio-fuels

Some early discussions of the impact of climate change suggested that rising temperatures would merely push crop farming farther North in the US and Europe. Thus farmers in Iowa and Nebraska might have to move to a warmer Canada to maintain crop production. Others suggested that increased carbon dioxide (CO_2), the main greenhouse gas, might *increase* crop production, since plants consume CO_2 while producing oxygen, just the opposite of homo sapiens. But Bourne, with a BS in agronomy from North Carolina State University cites detailed crop experiments at the University of Illinois which showed that any boost in crop yield due to higher CO_2 levels is more than offset by lower yields from higher temperatures. The Illinois researchers projected that corn and soy bean yields would *decrease* by 30–46 % under the Intergovernmental Panel On Climate Change's (IPCC) slowest warming scenario, and by 63–87 % under the IPCC's higher warming scenario.

1.1.1 Population Growth

The world's population grew from 3 billion in 1960 to 6 billion in 2000, the fastest doubling of population in world history. Fortunately, the green revolution with better seeds, fertilizers and pesticides, boosted grain production by a factor of 3 in this time period allowing sufficient crop production for most parts of the world. However, the UN projects a world population approaching 20.3 billion by 2100. Unless there is another version of the Green revolution over the coming decades which seems highly unlikely, it will be a strain to provide food for all these people.

1.1.2 Increased Meat Consumption

As many citizens of the world strive to meet the standard of living of the developed nations, they will undoubtedly aspire to eat more meat. But meat production requires much more crop production than eating grains, fruits and vegetables [2]:

Growing one pound of farmed salmon requires approximately one pound of grain,

Growing one pound of chicken requires 2 lb of grain,

Growing one pound of pork requires 3 lb of grain,

Growing one pound of beef requires 7 lb of grain,

As the citizens of other countries strive to increase their meat consumption to match the meat consumed by citizens of developed countries, more crops must be grown.

One UK study estimated that if the English cut their meat consumption in half, replacing meat with fruits, vegetables and grains, then their greenhouse gas emissions could be reduced by 19 %, while saving 43,600 lives each year in the process by reducing diet-related diseases [2].

1.1.3 Bio-Fuels

Producing bio-fuels such as bio-diesel and ethanol to power our motor vehicles already consumes approximately one third of all US corn production. One study estimated that by 2030 10 % of all US arable land might be devoted to producing bio-fuels, and 15 % of all land in the EU [2]. The impact of adding ethanol to gasoline does have three positive but relatively minor impacts: slightly reducing gasoline consumption, reducing carbon monoxide local air pollution, and modestly cutting greenhouse gas (GHG) emissions. Wang estimates that E-10, the mixture of 10 % ethanol and 90 % gasoline by volume that is currently blended in most regions of the US, reduces gasoline consumption by approximately 6.3 %[1] and E-10 reduces GHG emissions approximately 2 % [3].

While ethanol currently consumes one third of all US corn production, the ethanol industry is now trying to develop cellulosic ethanol plants that consume corn stover (the stalks and roots of the corn plant instead of the editable corn grain). Successful development of commercial cellulosic ethanol would eliminate the food vs fuel dilemma of using the corn grain to make ethanol. Cellulosic ethanol has the added advantage of greater GHG reductions since the ethanol is made from a waste product. Wang estimates that E-10 cellulosic ethanol reduces GHGs by 6 % versus a 3 % reduction for E-10 corn ethanol. Furthermore, if ethanol concentrations in gasoline could be increased to the 85 % level (E-85), then GHG emissions could be reduced by 64 % for cellulosic E-85 ethanol compared to only 17–23 % GHG reductions for E-85 corn ethanol depending on the ethanol plant milling process (wet vs. dry milling) [3].

Summary impacts of climate change. While the standard litany of climate change impacts (increased severe weather events, rising sea levels and global warming) are troubling, the prospect of not being able to feed the growing global population could be catastrophic. As Bourne states it:

> It's not hard to imagine an aging, wealthy and heavily armed Fortress Europe or Fortress America inhabited by 20 % of the world's temperate agricultural "haves" attempting to

[1]Ethanol has lower energy content than gasoline (76,330 btu/gallon for ethanol vs. 116,090 btu/gallon for gasoline) so adding 10 % ethanol to gasoline does not result in a 10 % decrease in gasoline consumption.

> wall off the remaining 80 % of the world's population—the young, poor, tropical and subtropical agricultural "have nots" who will do their best to get in [2]

This could make the 2014–2015 European immigration crisis from those fleeing wars in the Middle East seem mild by comparison to hundreds of millions of immigrants fleeing to seek adequate food for their families.

1.2 Fuel Sources

1.2.1 Coal as an Energy Source

To many advocates of clean energy, coal is an anathema: coal is the worst fossil fuel in terms of carbon content per unit energy.

As shown in Table 1.1, the average mixture of coal used for U.S. Electricity generation has 1.68 times more carbon content per MBTU of energy than natural gas, while lignite coal has 2.54 times more carbon than natural gas per unit energy, according to the Argonne National Laboratory GREET model [4, 5]. So burning coal to produce electricity would generate at least 68 % more CO_2 than burning natural gas to generate electricity, assuming equal efficiency for both coal and natural gas plants. But, in general, natural gas power plants are also more efficient than coal plants, meaning fewer MBTU's of natural gas are required to produce a given quantity of electricity than coal, which further increases coal GHG emissions compared to natural gas GHG emissions per kWh of electricity generated. By capturing the carbon as carbon dioxide (CO_2) at the electricity plant and storing it

Table 1.1 Carbon content of common fuels per unit energy and ratio of those values to natural gas

	g CO_2/MBTU	Ratio to natural gas
Gasoline blend stock	76,839	1.29
Low-sulfur diesel	79,072	1.33
Residual oil	85,081	1.43
Natural gas	59,413	1.00
Pure methane	58,028	0.98
Coal mix for electricity generation	100,000	1.68
Bituminous coal	89,913	1.51
Subbituminous coal	111,038	1.87
Lignite coal	151,141	2.54
Synthetic coal	118,476	1.99
Waste coal	109,166	1.84
Pet coke	98,336	1.66
Tire derived fuel	60,871	1.02
Coking coal	101,002	1.70

underground, however, coal can become a significant source of electricity while still meeting our carbon reduction goals to avert the worst consequences of climate change.

1.3 Hydrogen as an Energy Carrier for Fuel Cell Electric Vehicles

We have shown in a previous book that the hydrogen-powered fuel cell electric vehicle (FCEV) is the only alternative vehicle-fuel option that can approach our societal goal of reducing greenhouse gas (GHG) emissions to 80 % below 1990 levels in the light-duty transportation sector [6]. Large numbers of hybrid electric vehicles (HEVs) like the Toyota Prius could reduce GHGs by 37 % by the end of the 21st century, while plug-in hybrid electric vehicles (PHEVs) such as the Chevy Volt could reduce GHGs by 44 %. PHEVs running on a blend of 85 % ethanol in gasoline could reduce GHGs by 64 %, and battery electric vehicles (BEVs) could reduce GHG emissions by at most 70 % due to a combination of low market penetration[2] and the fact that most electricity in the U.S. today is generated by burning fossil fuels, primarily coal, which generates significant GHG emissions.[3] So society will need significant quantities of hydrogen and FCEVs to meet our GHG reduction goals in the transportation sector.

1.4 Hydrogen as an Enabler of Greater Use of Intermittent Renewables

Many analysts are advocating the use of renewable energy to provide the bulk of our electricity supplies in the future without any significant GHG emissions. But the two main renewables, wind and solar, are both intermittent. Any large scale dependence on these renewables will require other sources of backup energy when the wind is not blowing and the sun is not shining. Today electricity made from natural gas is the lowest carbon alternative to back up intermittent renewables. But electricity made from natural gas still generates significant GHG emissions, and would preclude our ability to achieve our GHG emission reduction goals.

Hydrogen made from coal with CCS could serve as a suitable backup for intermittent renewables, allowing society to eliminate the burning of fossil fuels for

[2]Low BEV market penetration is due primarily to the limited range and long refueling times of BEVs, combined with the difficulty of economically powering large vehicles such as trucks, buses and trains with batteries.

[3]One could even say that battery electric vehicles are really coal-burning vehicles in many regions in the U.S.!

our electric power needs. Hydrogen would be stored during sunny and windy periods. When wind and solar were not available, then the stored hydrogen would be used to provide electricity through a high efficiency stationary fuel cell system. Electrical efficiencies a high as 57 % are possible with fuel cells. If the excess heat from the fuel cell could be used in a combined heat and power (CHP) system,[4] then total efficiencies of 90 % or more are possible.

1.5 Hydrogen from Coal

While making hydrogen from renewables or nuclear power may be a worthy long-term goal, converting from our current electrical generation system based primarily on fossil fuels to one based on renewables and nuclear energy would be costly and disruptive. Many industries and countless thousands of jobs in the U.S. rely on coal today, including mining the coal and transporting it to electrical power plants around the nation. The coal industry is therefore a powerful economic and political force that might slow any sudden shift to renewable or nuclear hydrogen.

Making hydrogen from coal by conventional techniques such as burning the coal to raise steam to drive a gas turbine to make electricity to electrolyze water to produce hydrogen would be a losing proposition due both to cost and excessive GHG emissions.

There is a better option that has been commercialized at several power plants around the world. Rather than burning the coal to raise steam, the coal is first gasified at high temperature and pressure in a closed vessel. The resulting synthetic gas or "syngas" from the gasifier is composed primarily of hydrogen and carbon monoxide.[5] This syngas is then fed into a gas turbine to generate electricity; a heat exchanger then recovers some of the waste heat in this process to drive a steam turbine, producing more electricity in a coal-powered integrated combined cycle (IGCC) plant similar to natural gas combined cycle (NGCC) plants. In addition to producing electricity, the gasifier at the front end of the IGCC system can be oversized to produce more hydrogen than is needed for the electricity, so the IGCC plant can support both clean electricity and hydrogen for a near zero emission fuel cell electric vehicle (FCEV) transportation system.

Another alternative being developed today replaces the gas turbines with stationary fuel cell systems. The gasifier produces hydrogen as in the IGCC system, but this hydrogen is used to drive a high efficiency stationary fuel cell that produces electricity…an integrated gasification fuel cell system (IGFC).

---

[4]CHP systems will become more feasible with distributed power systems such as at large industrial complexes that can utilize both heat and electricity; or regional electricity generators might be coupled with district heating systems to more fully utilize the waste energy from a fuel cell system.

[5]This carbon monoxide (CO) is converted to more hydrogen and carbon dioxide in a water gas shift (WGS) reactor that combines CO and water or steam to form more hydrogen and CO_2: $CO + H_2O = CO_2 + H_2$.

1.6 Water Requirements for Hydrogen Generation

As described in Chap. 10, making hydrogen from coal requires less water than making hydrogen by electrolyzing water.

1.7 Water Requirements for Agriculture

More water will also be required for agriculture in a warmer world. At the present time, approximately 70 % of US fresh water resources are used in agriculture [2]. Approximately 20 % of irrigated farm land produces 40 % of the world's food supplies. Bourne reports that every 1 °C rise in temperature requires 3–4 % more water for irrigation [2]. In a warming world with growing population, then, more water will be required for grain production.

References

1. Mooney C, Warrick J (2016) 2015 is hottest year on record. The Washington Post, 21 Jan 2016
2. Bourne JK Jr (2015) The end of plenty: the race to feed a crowded world. W. W. Norton & Company. ISBN 978-0-393-07953-1
3. Wang M (2005) The Argonne National laboratory "Energy and greenhouse gas impacts of fuel ethanol." NGCA Renewable Fuels Forum, the National Press Club, 23 Aug 2005. Available at http://www.oregon.gov/energy/P-I/docs/WangonCorn050823.pdf
4. Wang MQ (2008) The Greenhouse gas, regulated emissions and energy use in transportation (GREET). Energy Systems Division, Argonne National Laboratory, available at http://greet.es.anl.gov/
5. Cai H, Wang M, Elgowainy A, Han J (2012) Updated greenhouse gas and criteria air pollutant emission fractions in their probability density functions for electric generating units. Argonne National Laboratory ANL/ESD/ 12-2 May 2012
6. Thomas CE (2015) Sustainable transportation options for the 21st century and beyond. Springer Publishing book in print, ISBN # 978-3-319-16831-9. Available online at: http://www.springer.com/978-331916-831-9

Chapter 2
Greenhouse Gases by Sector

Abstract Most analysis of greenhouse gas (GHG) emissions in the literature concentrates on motor vehicle emissions, since transportation generates the most GHG emissions at 33 % (with the industrial sector second at 27 %) if the GHGs from electricity are distributed to each end-use sector. But if electricity is counted as a separate "sector", then electricity in 2014 accounted for 39.4 % of all U.S. GHG emissions in 2013 (with transportation second at 33.7 %). Electricity is the major source of U.S. GHG emissions since most electricity is generated from fossil fuels, particularly coal, the highest carbon content fuel. The U.S. Energy Information Administration (EIA) projects that fossil fuels will still generate 61.6 % of all electricity by 2040, down only slightly from 64.9 % in 2015.

2.1 Greenhouse Gas Emissions by Sector

As shown in Fig. 2.1, if we consider the four sectors of society (Commercial, Industrial, Residential and Transportation) with the GHG emissions from electricity generation assigned to each end-use sector, then transportation accounts for the largest share of GHG emissions at 33.8 %, with the industrial sector generating the next highest percentage at 27 % according to the U.S [1].

However, Fig. 2.1 is based on EPA data with the GHGs from electricity generation allocated to each sector where the electricity is consumed.

If we remove the electricity GHGs from each sector and include electricity as a separate GHG category, then electricity produces the largest share of GHGs (39.4 %), with transportation second at 33.7 % as shown in Fig. 2.2.

C.E. Thomas, *Stopping Climate Change: the Case for Hydrogen and Coal*,
Lecture Notes in Energy 35, DOI 10.1007/978-3-319-31655-0_2

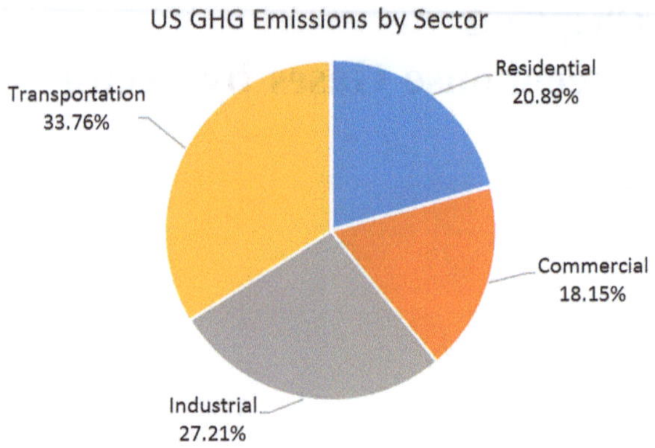

Fig. 2.1 U.S. greenhouse gas emissions in 2014 with GHGs from electricity distributed to each sector where it is consumed

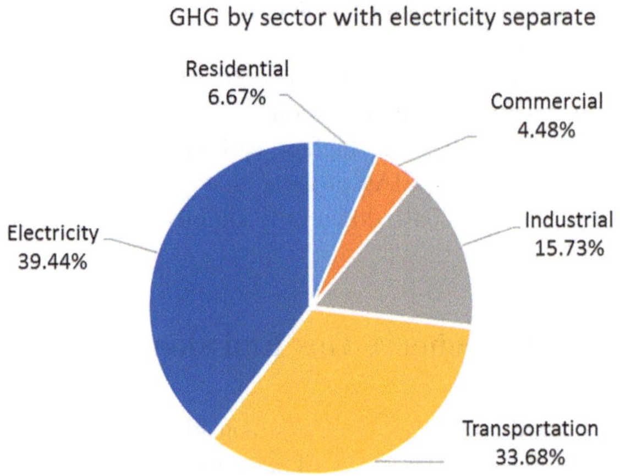

Fig. 2.2 U.S. GHG emissions in 2014 with emissions from electricity GHGs shown separately

2.2 Electricity Projections

The EIA's projection of U.S. electricity generation by fuel though 2040 is shown in Fig. 2.3. Natural gas and diesel fuel use increase slightly over this period, while coal use decreases. Renewable energy rises slightly by 2020, with a slower but steady rise after 2020 according to EIA projections. The net fossil fuel used for generating electricity is fairly flat through 2040 according to the EIA projections in their Annual Energy Outlook (AEO) for 2015 [1]. Fossil fuels produce 64.9 % of all U.S. electricity in 2015, falling only slightly to 61.6 % by 2040 according to the EIA projections in their reference case. This AEO "reference case" assumes that the U.S. Congress passes no legislation to cut greenhouse gas emissions, and is therefore a "business-as-usual" scenario.

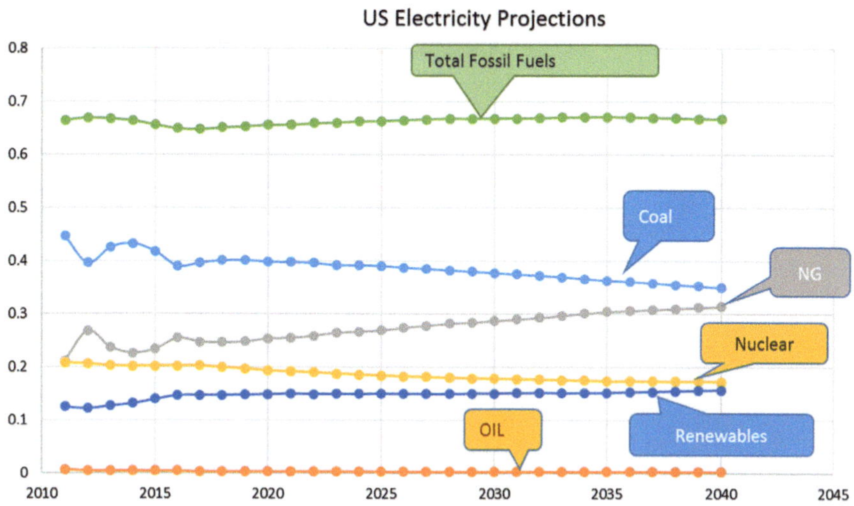

Fig. 2.3 Electricity fuel sources projected by the Department of Energy's 2015 "Annual Energy Outlook"

We conclude that we must replace existing electricity generators with lower GHG emitting power plants if we are to reduce electricity GHGs that accounted for 39 % of all U.S. emissions in 2015.

Reference

1. "Inventory of US greenhouse gas Emissions and Sinks-1990–2014" The U.S. Environmental Protection Agency, EPA # 430-R-16-002, April 15, 2015. Available at: https://www3.epa.gov/climatechange/Downloads/ghgemissions/US-GHG-Inventory-2016-Main-Text.pdf

Chapter 3
Net Greenhouse Gas Reductions Required by Sector

Abstract The climate change community has called for an 80 % reduction in greenhouse gas (GHG) emissions below 1990 emission levels, preferably by 2050 to avoid the worst consequences of climate change in the 21st century. In this chapter we analyze the required reductions in each sector of U.S. society to obtain an overall 80 % reduction below 1990 levels. This assessment suggests that the GHG emissions from generating electricity should be reduced to at least 58 % below 1990 electricity emission levels, or a reduction from 1866 million tonnes of CO_2-equivalent in 1990 to less than 784 million tonnes of GHG emissions by 2050.

3.1 U.S. Greenhouse Gas Emissions in 1990

Table 3.1 lists the U.S. 1990 greenhouse gas (GHG) emissions in the second and third columns, based on the U.S. Environmental Protection Agency (EPA) GHG data [1]. The second column lists the GHG emissions with the electricity GHG emissions allocated to the five main sectors (Agriculture, Residential, Commercial, Industrial and Transportation). The third column of Table 3.1 removes the electricity portion of GHG emissions from these five sectors, and assigns them to a separate electricity sector. In 1990, electricity production accounted for 34.8 % of all U.S. Greenhouse gas emissions. The "LDV" transport sector includes all light duty vehicles (LDVs) including passenger vehicles, station wagons, trucks, vans, SUVs, etc. The "other on-road transport" sector includes all medium and heavy duty trucks and buses. The "Other transport" row includes all ships, planes, trains, as well as emissions from pipelines as shown in Table 3.2. Two companies have announced plans to supply hydrogen-powered fuel cells to drive trains [4, 5].

C.E. Thomas, *Stopping Climate Change: the Case for Hydrogen and Coal*,
Lecture Notes in Energy 35, DOI 10.1007/978-3-319-31655-0_3

Table 3.1 1990 GHG emissions by sector (in million metric tonnes of CO_2-equivalent) from the EPA greenhouse gas inventory [1]

	1990 GHG levels including electricity in sector	1990 GHGs by sector (electricity separate)	1990 GHG %'s
Residential	953.1	345.4	6.4
Commercial	936.7	385.3	7.2
Industrial	2173.9	1531.5	28.5
Transportation	1556.3	1553.2	28.9
Electricity	–	1866.1	34.8
Agriculture	−251.7	−313.0	−5.8
Total GHGS	5368.3	5368.5	100.0

Table 3.2 1990 GHG emissions from transportation sectors in million tonnes of CO_2-equivalent emissions per year

	1990	
	Mtonnes	%
Passenger cars	657.4	42.2
LD trucks	336.6	21.6
Buses	8.4	0.5
Med&HD trucks	231.1	14.8
Commercial aircraft	110.9	7.1
Other aircraft	78.3	5.0
Rail	39	2.5
Ships and boats	45.1	2.9
Pipelines	36	2.3
Other	13.6	0.9
Totals	1556.4	

3.2 Recommended GHG Reductions by Sector

An 80 % reduction in 1990 GHG emissions for the U.S. would require total emissions of less than 1074 million tonnes of CO_2-equivalent. We have previously shown that if the light duty vehicle sector shifts primarily to hydrogen-powered fuel cell electric vehicles (FCEVs), then the transportation sector GHG reductions could be reduced to 95.7 % below 1990 levels by 2100 or 54 million tonnes [2]. The net agriculture GHG emissions include both the net agriculture emissions minus the GHG sinks from agriculture and forest growth in vegetation that absorbs CO_2 from the atmosphere as summarized in Table 3.3 for 1990 and 2012 from the latest GHG inventory from the EPA[1] [3]. We use the 2012 estimate of minus 365.5 million tonnes for the agriculture sector.

[1]The latest year available from the EPA inventory was 2012.

Table 3.3 Agriculture sources and sinks for GHG emission in million tonnes of CO_2

	Source	Sink	Net
1990	518.1	−831.1	−313
2012	614.1	−979.3	−365.2

If we assume (arbitrarily) that Residential, Commercial, and industrial sectors all reduce 1990 emissions by 75 %, and we assume that the non-LDV transport sectors could reduce their emissions by 88–92 %, then GHG emissions from generating electricity would have to be reduced by 58 % below 1990 levels or 784 million tonnes per year to achieve our over-all goal of an 80 % decrease in GHGs from 1990. These reductions would result in an overall net reduction of just over our goal of an 80 % reduction for all U.S. GHG emissions as shown by the bottom line of Table 3.4.

If any of the other sectors (Residential, Commercial or Industrial) are unable to achieve a 75 % reduction in GHG emissions as assumed in Table 3.4, then the electricity sector might need to reduce GHG emissions more than 58 %. A 75 % reduction in residential emissions might be difficult, due to the slow turnover in housing stock; in essence, the residential sector reductions would have to be achieved primarily through improved energy efficiency of household appliances, lighting, extra insulation, etc.

In addition, it might be difficult for the industrial and commercial sectors to achieve a 75 % reduction below 1990 levels, since these sectors have already made

Table 3.4 Suggested GHG reduction goals for the five major sectors, which establishes a reduction goal of 58 % below 1990 levels for the electricity generation sector to meet the overall societal goal of an 80 % reduction below 1990 levels

	1990 GHG levels including electricity in sector	1990 GHGs by sector (electricity separate)	1990 GHG %'s	Suggested % reductions below 1990	Net GHG goals (Mtonnes CO_2)
Residential	953.1	345.4	6.4	75.0	86.4
Commercial	936.7	385.3	7.2	75.0	96.3
Industrial	2173.9	1531.5	28.5	75.0	382.9
LDVs	997.3	994.0	18.5	97.1	28.8
Other on-road transport	239.5	239.5	4.5	92.0	19.2
Other transport	322.9	322.9	6.0	88.0	38.7
Electricity	–	1866.1	34.7	58.0	783.8
Agriculture	−251.7	−313.0	−5.8	−16.7	(365.2)
Total GHGS	5371.7	5371.7			1070.8
Net % reduction in total GHG emissions below 1990 levels					−80.1 %

Table 3.5 GHG emissions by sector in 2015 and percentage reductions relative to 1990

Million tonnes CO_2	1990 GHGs by sector (electricity separate)	2015 sector emissions (electricity separate)	2015 change w/r to 1990 levels (%)
Residential	345.4	312	−9.7
Commercial	385.3	225	−41.5
Industrial	1531.5	1030	−32.7
Transportation	1553.2	1807	16.3
Electricity	1866.1	2054	10.1
Agriculture	−313.0	−365.5	16.7
Total GHGS	5368.5	5062.6	−5.7

large reductions in their emissions. As shown in Table 3.5, Commercial GHG emissions were already 41.5 % below 1990 levels in 2015, while industrial emissions were 32.7 % below 1990 levels. It might be difficult and/or costly for these two sectors to cut GHG emissions further.

Likewise, it may be difficult to reduce emissions from heavy duty vehicles, ships, planes and trains[2] by 92 % as assumed in Table 3.4. In that case the GHG emissions from generating electricity would have to be cut more than the 58 % decrease derived in Table 3.4.

We show in Chap. 4 that GHG emissions from the electricity sector could be cut by more than 83 %, or greater than the 58 % cut assumed in Table 3.4. In this case, we could reduce the GHG targets for other sectors. For example, Table 4.4 in Chap. 4 shows one combination of reductions, assuming the 83.6 % reduction of electricity GHGs that is feasible; this would permit us to reduce the reduction target for the industrial sector from the 75 % reduction assumed here to only 60 % below 1990 levels, and residential and commercial reduction targets could both be reduced from 57 to 70 %. Other combinations are possible, since the increased GHG reductions from the electricity sector postulated in Chap. 4 could be distributed to other sectors that cannot meet their reduction goals.[3]

There is one caveat, however, with these reduced electricity GHGs. The 83.6 % electricity reduction target discussed above is achieved by replacing some coal-based IGCC + CS power plants with NGCC + CCS power plants, which increases the consumption of natural gas. As shown in Chap. 5, this would put a strain on U.S. natural gas reserves. Even if we assume the natural gas TRR (total recoverable reserves) which is much larger than natural gas *proved* reserves, the

[2]Although Ballard signed a contract in November 2015 with a Chinese company to supply 200-kW fuel cell modules to power trams in China, and Hydrogenics signed a contract with Alstom Transport to produce fuel cells for trains in Europe, which would significantly reduce the emissions from rail traffic that now depends on high carbon diesel fuel.

[3]It may also be less costly to make reductions in other sectors than in the industrial or electricity sectors.

total natural gas reserves would be consumed by 2089 under this minimum electricity GHG case. For comparison, the total accessible[4] U.S. coal reserves would last until 2258 under this scenario.

References

1. Inventory of U.S. greenhouse gas emissions and sinks: 1990–2012, The U.S. Environmental Protection Agency, EPA 430-R-14-003, 15 Apr 2014. Available at http://www.epa.gov/climatechange/Downloads/ghgemissions/US-GHG-Inventory-2014-Main-Text.pdf
2. Thomas (CE) (2015) Sustainable transportation options for the 21st century and beyond: a comprehensive comparison of alternatives to the internal combustion engine. Springer International Publishing. ISBN# 978-3-319-16832-9
3. "Inventory of U.S. greenhouse gas emissions and sinks: 1990–2012", the U.S. Environmental Protection Agency, EPA 430-R-14-003, 15 Apr 2014. Available at http://www.epa.gov/climatechange/Downloads/ghgemissions/US-GHG-Inventory-2014-Main-Text.pdf
4. Ballard signs $3 M contract for development of fuel cell module to power trams in China, 1 Nov 2015. Available at http://www.ballard.com/about-ballard/newsroom/news-releases/news11011501.aspx
5. Hydrogenics and Alstom transport sign agreement to develop and commercialize hydrogen-powered commuter trains in Europe, 27 May 2015. Available at http://www.hydrogenics.com/about-the-company/news-updates/2015/05/27/hydrogenics-and-alstom-transport-sign-agreement-to-develop-and-commercialize-hydrogen-powered-commuter-trains-in-europe

[4]We use the estimate from the EIA that 53 % of known coal reserves are accessible.

Chapter 4
Electricity Generator Costs and Greenhouse Gas Emissions

Abstract The capital costs and greenhouse gas (GHG) emissions from various types of electricity power plants will determine the least cost option for achieving our goal of reducing GHG emissions by at least 58 % below 1990 levels from the electricity generation sector (as a major part of our goal of reducing all GHG emissions by 80 % below 1990 levels by 2050). We show a plausible, economic scenario whereby replacing current fossil fuel based power generators with coal-powered integrated gasification combined cycle plants (IGCC) with carbon capture and storage (CCS) will achieve our GHG reduction targets by 2050.

4.1 Introduction

Without any incentive to cut GHGs such as a carbon tax, carbon fee or "Cap and Trade" legislation, the utility industry in the U.S. will most likely continue with its current gradual replacement of coal-powered generators with higher efficiency natural gas combined cycle (NGCC) power plants to respond to clean air regulations; while this trend will gradually reduce GHG emissions, it will fall far short of achieving an overall 80 % reduction below 1990 levels by 2050, the goal of the climate-change community. In this chapter we explore the GHG emissions and capital cost of four options:

- "Business as usual" based on a continuation of the current trend of gradually replacing coal with natural gas
- Lower carbon options including increased energy efficiency, increased renewable energy and increased nuclear energy
- Coal-based integrated gasification combined cycle (IGCC) plants with carbon capture and storage/sequestration (CCS), and
- Natural gas-based combined cycle (NGCC) plants with CCS.

In the following sections, after reviewing the estimated capital costs of alternative electricity generators, we estimate the GHG reductions and capital costs of each of these four pathways.

C.E. Thomas, *Stopping Climate Change: the Case for Hydrogen and Coal*,
Lecture Notes in Energy 35, DOI 10.1007/978-3-319-31655-0_4

4.2 Business as Usual Pathway

In the business-as-usual (BAU) pathway, we assume that the utilities continue on their current pathway of gradually replacing older coal-powered steam generators with natural gas-powered combined cycle (NGCC) power plants, while adding some renewable energy sources as mandated in many states. Even if we replaced *all* fossil fuel plants with NGCC plants by 2050, we could not meet our goal derived in Chap. 3 that electricity should generate less than 784 Million tonnes of CO_2-eq. by 2050. Even If *all* the estimated 712 GW of fossil electricity capacity were provided by NGCC plants in 2050, then these plants would still generate 3005 Million tonnes[1] of CO_2-eq. per year or 3.8 *times* the 784 Million tonne goal.

4.3 Lower Carbon Options

There are three primary approaches to reducing greenhouse gas (GHG) emissions from electricity power plants:

- Change fuels
- Add carbon capture and storage/Sequestration (CCS)
- Increase use of nuclear and renewable energy

4.3.1 Change Fuels

GHG emissions per kWh of electrical energy can be reduced by switching from fuels with high carbon content such as coal and petroleum products to lower carbon fuels such as natural gas or particularly zero-carbon renewable fuels such as the sun, wind, hydroelectricity, geothermal or wave power, or nuclear power. The average US coal mix used for electricity generation has 68 % more carbon than natural gas per unit energy, and lignite coal has 2.54 times more carbon per unit energy. Converting from coal to natural gas power plants therefore reduces GHGs.

The measured or estimated HHV efficiencies[2] of various existing or planned electrical generators are shown in Table 4.1 from three sources [1–3].

[1]Assuming a 95 % plant capacity factor and NGCC plants emitting 507 g of CO_2-eq. per kWh of electricity (Table 4.2).

[2]Higher heating value (HHV) efficiencies are used here since the utility fuels (coal and natural gas) are price in $/MBTU on a higher heating value basis.

Table 4.1 Estimated higher heating value (HHV) power plant efficiencies

	HHV heat rate			Ave. HHV	HHV Eff
	IEA-2011	EIA-2015	EPRI-2015	Heat rate	Average (%)
Coal-ST	7763	8740	11,145	9216	37.0
IGCC	8749	7450	8900	8366	40.8
IGCC + CCS	10,943	8307	10,250	9833	34.7
NGCC	6400	6333	7000	6578	51.9
NGCC + CCS	7484	7493	7590	7522	45.4

Table 4.2 Average costs ($/kW), greenhouse gases and efficiencies of NGCC and IGCC plants with and without CCS

	IGCC	IGCC + CCS	CCS change (%)
Efficiency (% HHV)	40.8	34.7	85.1
Capex ($/kW)	$2866	$4207	146.8
GHGs (g/kWh)	1112	87.4	7.9
	NGCC	NGCC + CCS	CCS change (%)
Efficiency (% HHV)	51.9	45.4	87.4
Capex ($/kW)	$1003	$1811	180.5
GHGs (g/kWh)	507	50.9	10.0

4.3.2 Carbon Capture and Storage/Sequestration (CCS)

The GHG emissions from power plants can be substantially reduced by capturing the CO_2 at the plant and transporting it to an underground storage cavity- the carbon capture and storage/sequestration (CCS) option. The power plant efficiencies, capital costs and greenhouse gas emissions are summarized in Table 4.2 for both natural gas NGCC plants and coal-powered IGGC plants, both with and without CCS. Table 4.2 shows that the cost of an NGCC plant increases proportionately more by adding CCS than an IGCC plant (180 vs. 147 %).

The estimated capital costs in Table 4.2 are the average of five sources from the literature as summarized in Table 4.3.

The sources include:

- IEA-2011. A 2011 report from the International Energy Agency [1]
- EIA-2015. 2015 data from the U.S. Energy Information Administration [2]
- EPRI-2015 from the Electric Power Research Institute [3]
- WEIO-2020. A 2014 report[3] from the International Energy Agency [4]
- A 2007 National Energy Technology Laboratory (NETL) report [5]

[3]The "2020" in WEIO-2020 refers to the year for cost estimates, not the year of publication.

Table 4.3 Estimated costs of power plants from five literature sources in 2015 $/kW

	IEA-2011	AEO-2015	EPRI-2015	WEIO-2020	NETL-2007	Averages
Coal-ST	$2285	$2917	$2433	$2123	$1641	$2280
IGCC	$2733	$3747	$3287	$2628	$1934	$2866
IGCC + CCS	$3925	$6527	$4152	$3811	$2622	$4207
NGCC	$1014	$1022	$1084	$1314	$582	$1003
NGCC + CCS	$1812	$2083	$2110	$1820	$1231	$1811
Nuclear	ND	$5414	$5791	$5054	ND	$5420

All costs have been converted to 2015 dollars using the Power Capital Cost Index (PCCI) to account for inflation in the power industry since these reports were written [6]. The GHG emissions listed in Table 4.2 are taken from the Argonne National Laboratory's "well-to-wheels" GREET model [7].

One prime example of reducing GHGs with CCS is replacing existing steam boiler coal power plants with coal-powered integrated gasification combined cycle (IGCC) plants, including carbon capture and storage/sequestration (CCS). IGCC plants are still powered by coal, but instead of burning the coal to produce steam to run a turbine that sends the carbon from the coal up the chimney as carbon dioxide (CO_2), the coal is gasified at high pressure and temperature in an enclosed chamber to produce a synthetic gas or "syngas" made up primarily of hydrogen and carbon monoxide. The carbon monoxide can be converted to more hydrogen by the water gas shift reaction ($CO + H_2O = CO_2 + H_2$). The CO_2 and local air pollution gases can be removed from this intermediate gas stream more readily than trying to remove the gases from the hot exhaust gases from conventional power plants. The resulting hydrogen can be burned cleanly in a turbine connected to a generator to produce electricity. The hydrogen can also be used to power fuel cell electric vehicles (FCEVs), and it can be used to produce electricity in a stationary fuel cell system with even higher efficiency. The carbon dioxide from the IGGC plant is more concentrated than the CO_2 in the exhaust from a steam turbine, so removing the CO_2 prior to combustion is easier in an IGCC plant. Thus a coal-powered IGCC system with CCS can support both low carbon electricity and very low carbon transportation via fuel cell electric vehicles (FCEVs), thereby addressing the two greatest sources of GHGs.

Note that the estimated cost for IGCC + CCS plants in Table 4.3 is low compared to current experience. For example, the initial US IGCC + CCS plant in Kemper County, Mississippi, was originally estimated to cost $2.2 billion, but now has surged to a cost projection of $6.6 billion [8] for 582 MW of power, or $11,340/kWe, or more than 2.6 times the average estimated cost of $4207/kWe for IGCC + CCS plants from Table 4.3. According to one source regarding the Kemper plant [9].

> The plant has suffered almost every kind of cost overrun, beset by bad weather, labor costs, shortages and "inconsistent" quality of equipment and materials, and contractor and supplier delays.

The CEO of the Southern Company that runs the Kemper plant stated that

> "we did not evaluate correctly the amount of pipe, the thickness of the pipe, the metallurgy of the pipe and the quantity of the pipe," he said. "When all that was fixed, the support structures had to be made stronger, too." [9]

While the IGCC + CCS plant at Kemper came in at a much higher cost than the average cost projected by the experts, the next plants should gain from the experience of the first plant. At the very least the next plants built should not be plagued with errors in pipe calculations and the need to replace under-sized pipes. In this book we therefore assume that the average cost of the next generation of IGCC + CCS plants will achieve the $4207/kWe listed in Table 4.3. Even if initial costs are higher than this estimate, given the environmental and national security advantages of these plants to society as documented in other chapters of this book, we can anticipate that governments might provide incentives to the builders of the next generation of low carbon fossil fuel generators such as IGCC + CCS plants, just as they provide incentives now for low or zero carbon renewable electricity plants.

The costs for two IGCC plants in the US without CCS are closer to the literature average. The 262-MW Wabash Power plant in Indiana cost $438 million in 2000$, or approximately $769 million in 2015$,[4] or a cost per kW of $2993/kW. The 250-MW Polk IGCC plant in Florida cost $448 million in 2000$, or approximately $786 million in 2015$, or a cost of $3144/kW [10]. However, Tampa Electric estimates that if they built another plant, it would cost only $412 million in 2000$ due to lessons learned with the first plant, or $723 million in 2015$, or a cost of $2892/kW. The average of the Wabash plant cost and the future estimated cost of the Polk plant is $2913/kW, or less than 2 % above the literature average of $2886/kW for an IGCC plant from Table 4.3.

4.4 Total Power Plant Costs to Meet GHG Goals

From Chap. 3, we set a goal of reducing GHG emissions from electricity generation by 58 % below 1990 levels[5] or a goal of generating less than 784 million metric tonnes of CO_2-equivalent per year from electricity production. To reduce GHG emissions, we make the following assumptions about the US power grid over the 21st century:

---

[4]The Power Capital Cost Index (PCCI) from 2001 to 2015 was a factor of 1.755.

[5]As part of a plan to reduce all U.S. GHG emissions by 80 % below 1990 levels.

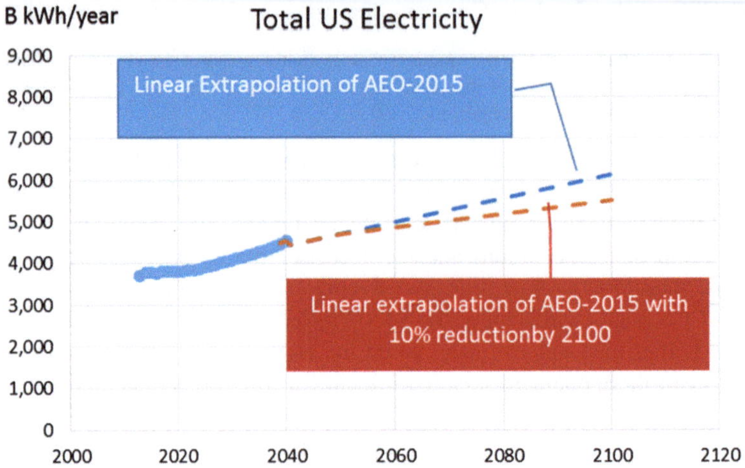

Fig. 4.1 Total U.S. electricity generation with a 10 % reduction in the linear extrapolation of the AEO-2015 projections through 2040

4.4.1 Energy Efficiency Gains

First, we assume an across-the-board 10 % improvement in the efficiency of using electricity across all sectors of the economy after 2040. We apply this efficiency increase to the linear extrapolation of the Energy Information Administration's 2015 Annual Energy Outlook beyond 2040 as shown in Fig. 4.1. By 2100, the total electricity production drops from 6128 billion kWh/year for the linear extrapolation to 5515 B kWh/year with our assumed 10 % efficiency improvements.

4.4.2 Renewable Electricity Increases

For renewable electricity generation, we assume a 30 % *increase* over the AEO-2015 linear extrapolation of renewables from 2020 to 2100, as shown in Fig. 4.2. With this 30 % increase, the total renewable electricity in 2100 increases from 1463 B kWh/year for the AEO-2015 extrapolation to 1981 B kWh/year. We assume a linear ramp multiplier times the AEO-2015 projections of increased renewables from 2020 to 2050, starting with a multiplier of one in 2020 and increasing to 1.3 by 2100.

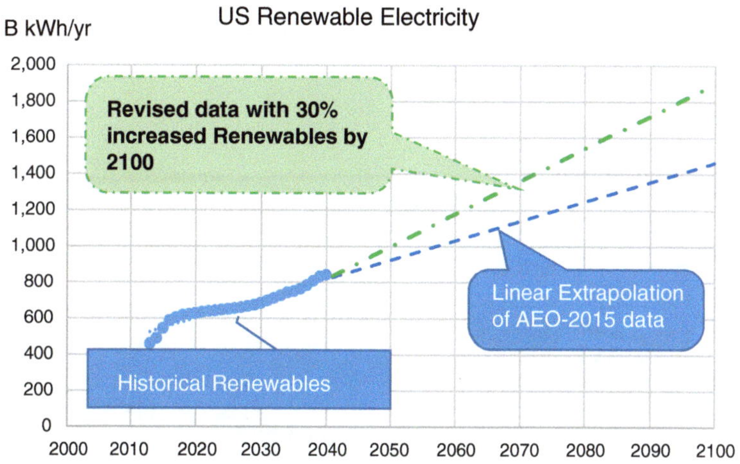

Fig. 4.2 Renewable electricity generation projections

4.5 Nuclear Electricity Projections

For nuclear energy, we assume a linear extrapolation of the AEO-2015 projections with no change, as shown in Fig. 4.3. Nuclear electricity grows modestly from 789 B kWh in 2015 to 961 B kWh in 2100 with the linear extrapolation of the AEO-2015 data

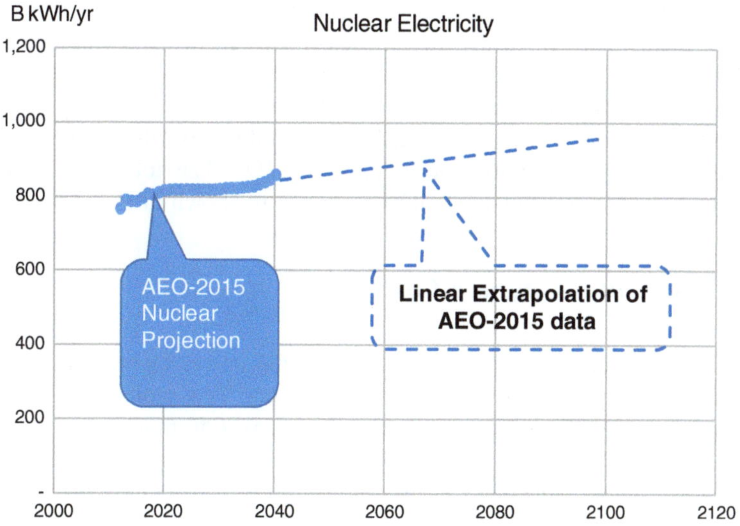

Fig. 4.3 US nuclear electricity projections

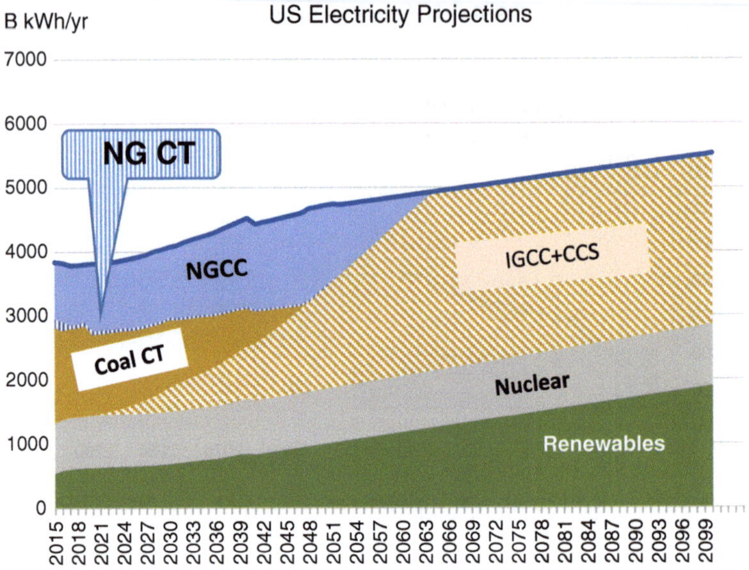

Fig. 4.4 Total electricity generation by generator type over the 21st Century to meet GHG goal by 2050

4.6 Generation Mix to Meet GHG Goal by 2050

One combination of power plants that meets our initial GHG reduction target of less than 784 million tonnes of CO_2-equivalent by 2050 is shown in Fig. 4.4. In this scenario, all coal, natural gas and oil combustion turbines are eliminated by 2048. These generators are replaced over the century by natural gas combined cycle (NGCC) plants and by coal-powered integrated gasification combined cycle (IGCC) plants with carbon capture and storage/sequestration (CCS). This combination of power plants reduces GHGs to 781 million tonnes by 2050 (or slightly less than our goal of 784 million tonnes by 2050), falling to 238 million tonnes/year by 2080, and to 332 million tonnes/year by 2100. The total electricity GHG emissions by source for this goal GHG scenario are shown in Fig. 4.5

4.6.1 Generator Upgrade Costs

The total annual capital expenses for building the power plants required for the generation mix of Fig. 4.4 are shown in Fig. 4.6. The capital costs are large, reaching just under $72 billion per year in the 2054–2056 time period; these costs include all the generators shown in Fig. 4.4, including nuclear and renewable energy power plants. Note that the costs shown in Fig. 4.6 are the total costs of *new* power plants; these costs do not account for the fact that the utilities would have to replace aging power plants over these years; a more appropriate cost metric might

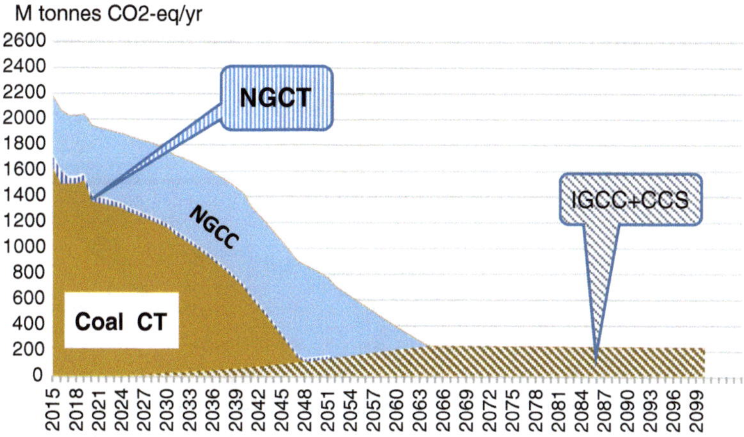

Fig. 4.5 Electricity GHG emissions by generator type for the Goal case of less than 784 Million tonnes CO_2-eq. by 2050

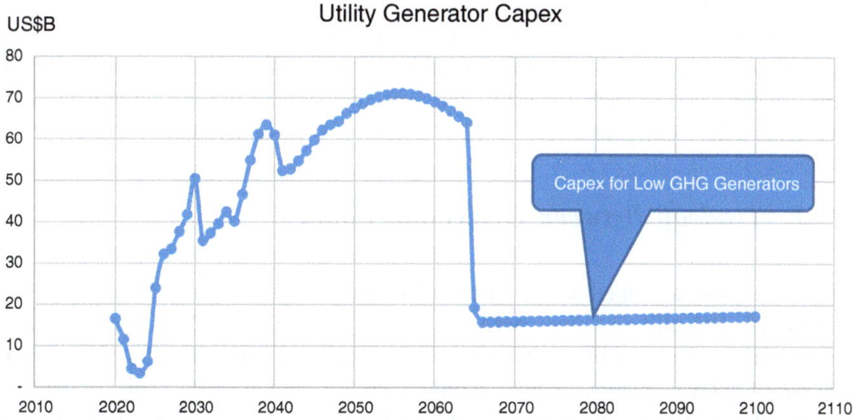

Fig. 4.6 Estimated utility capital investments necessary to achieve the carbon reduction levels shown in Fig. 4.5

be the *difference* between the cost of these low-carbon generators and the required replacement generators. After 2064 the only new plants are renewables and nuclear, both of which are assumed to increase gradually as described above.

4.6.2 Comparison Between Upgrade Costs and Historical Capital Expenditures

While large, these costs are not excessive compared to historical utility industry spending on capital equipment as demonstrated in Fig. 4.7. The total utility industry

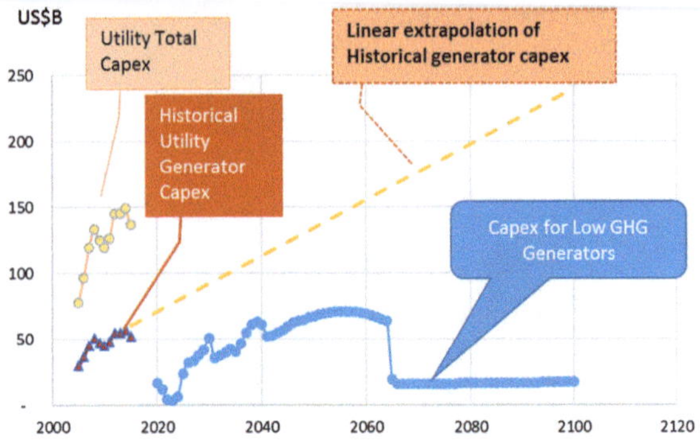

Fig. 4.7 Utility generator costs for the Goal GHG case compared to historical utility capital expenses

capital expenditures reached $149.6 billion in 2014; however, as discussed in Chap. 6, generation accounts for only 37 % of total IOU investments,[6] or approximately $55 billion for 2014. Thus the total investor owned utility (IOU) investments of $149.6 billion exceeds the estimated peak cost of $76 billion in 2054 for our postulated low-carbon grid that meets our societal goal of an 80 % reduction below 1990 GHG emissions. This peak spending only modestly exceeds the estimated utility generator spending that averaged $54.3 billion per year in 2012 through 2014. Also shown in Fig. 4.7 is a linear extrapolation of the historical trend in generator capex, which also exceeds the estimated peak in spending to reduce the grid GHG emissions below our goal of 784 million tonnes of CO_2-equivalent by 2050.

4.6.3 Tradeoff Between GHGs and Capex

While the costs described above would allow us to meet our goal of cutting GHGs to less than 784 million tonnes per year of CO_2-equivalent by 2050, we could also reduce GHG emissions *more* than this goal by building more IGCC + CCS plants instead of NGCC plants,[7] which would increase capital costs. This tradeoff is illustrated in Fig. 4.8. The independent variable in Fig. 4.8 is "A" the multiplier in the logistics function that defines the rate of introduction of IGCC + CCS plants.

A = 1.51 for the previous figures, which reduced GHGs slightly below our goal of 784 million tonnes by 2050. But if we increase A above 1.5, then more IGCC + CSS plants are built, which reduces GHG emissions as shown in Fig. 4.8. Thus we could reduce GHG emissions from the electricity sector down to approximately 530 million tonnes/year of GHGs by 2050, but then the maximum yearly capital costs for generators would increase from $58 billion up to $73 billion. So we have the flexibility to decrease electricity GHGs below our goal, which could help the other sectors of the economy if they fail to reach their goals of cutting GHGs to 75 % below 1990 levels.

[6]The other expenditures are for transmission and distribution of electricity.

[7]Alternately, we could add CSS to the NGCC plants as discussed in the Sect. 4.7 below.

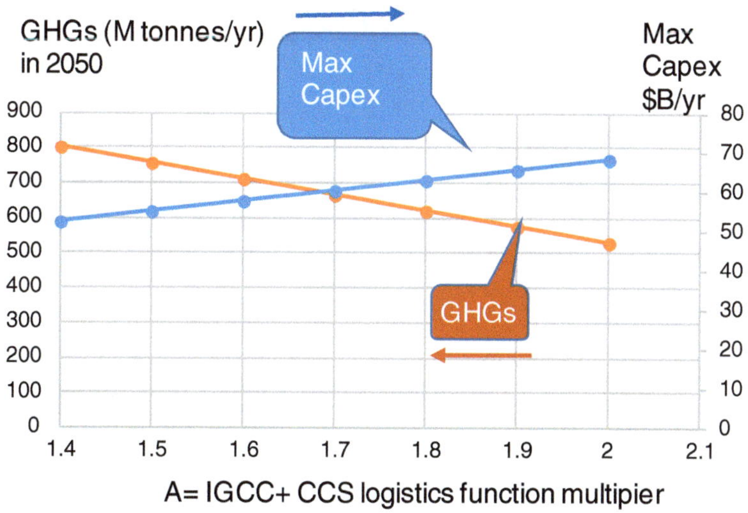

Fig. 4.8 Tradeoff between 2050 GHG emissions and electricity generator maximum annual capital expenditure

4.7 Scenario with Minimal GHGs (NGCC + CCS Plants)

Another option to further reduce GHGs would be to add carbon capture and storage/sequestration (CCS) to natural gas combined cycle (NGCC) plants. These NGCC + CSS plants are estimated to have approximately 43 % of the cost of the IGCC + CCS plants and to generate 42 % lower GHGs as shown in Table 4.2 above. One combination of generators that reduces GHGs to less than 306 million tonnes of CO_2-equivalent by 2050 is shown in Fig. 4.9. The resulting GHG values are shown in Fig. 4.10. GHG emissions fall to 191 million tonnes per year by 2070 and to only 184 million tonnes per year by 2100.

The annual capital expenditures for this minimum GHG scenario are shown in Fig. 4.11. In this case the maximum annual expenditures are approximately $79 billion in 2051, more than the maximum of $71 billion per year required for the GHG goal case scenario. As shown in Fig. 4.12, these expenditures are not excessive compared to historical utility capital expenditures.

4.8 Scenario with GHG Goal Delayed to 2070

The scenario to meet the GHG goal by 2050 described in Sect. 4.6 above would require the introduction of IGCC + CCS plants by 2020 to 2030 as shown in Fig. 4.4. This is may be doubtful in the U.S., since the further development and

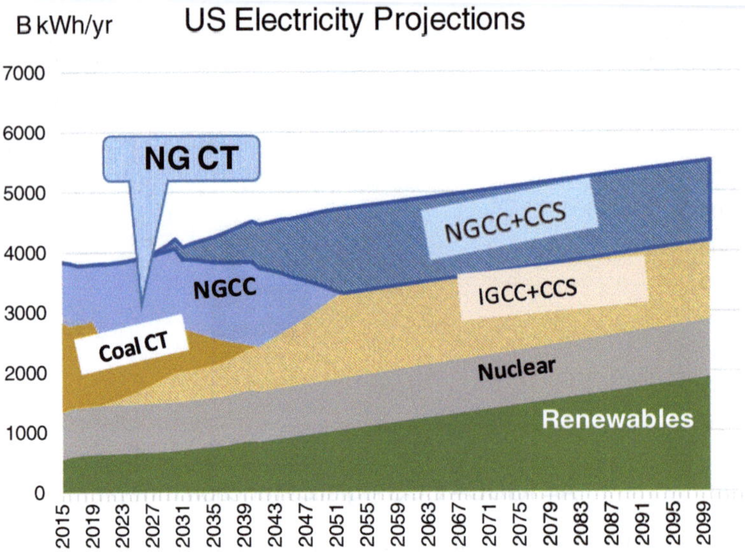

Fig. 4.9 Electricity generation mix with NGCC + CCS plants to reduce GHG emissions to less than 306 million tonnes per year by 2050

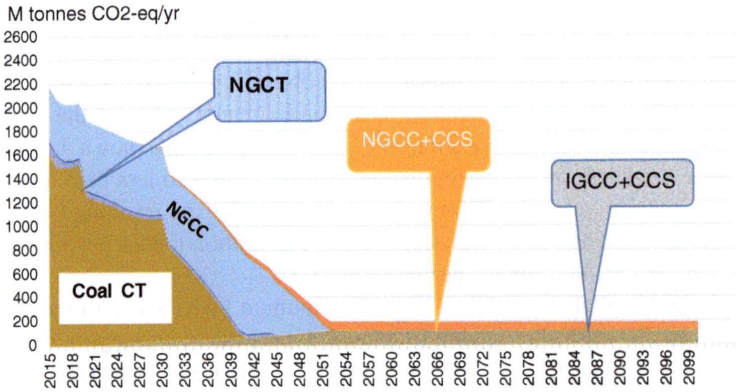

Fig. 4.10 Minimum GHG scenario with NGCC + CCS plants, with less than 305 million tonnes of CO_2-eq. by 2050

introduction of IGCC + CCS plants would most likely require strong Federal action, which is very unlikely, given that many members of the current U.S. Congress are "climate change deniers" who would not approve extra investment in lower carbon generators or any significant carbon fee or tax. But even if the GHG goals are delayed to after 2050, the requirements of deploying low-carbon

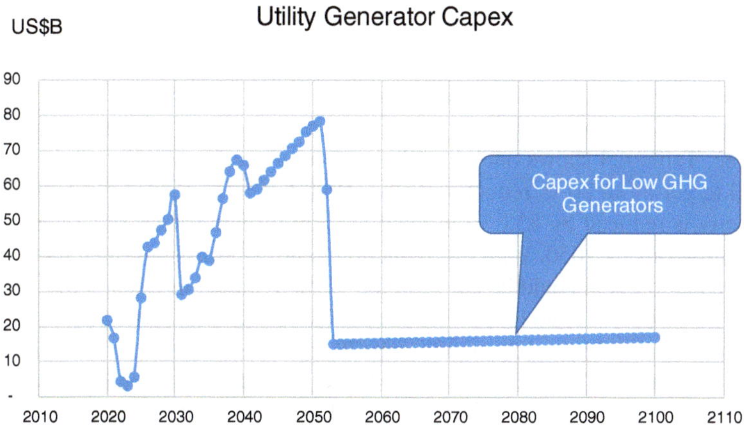

Fig. 4.11 Annual capital expenditures to install the generators for the minimum GHG goal case of Fig. 4.9

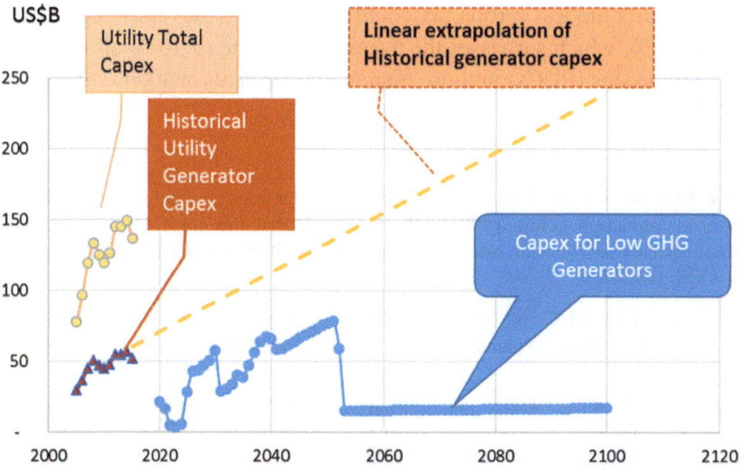

Fig. 4.12 Capex for min GHG case compared to historical utility expenditures

electricity generators with CCS would remain. By delaying the introduction of IGCC + CCS plants by 20 years, GHG's could be reduced to 241 million tonnes of CO_2-equivalent by 2070, well below our initial goal of 784 million tonnes per year. However, the maximum annual capital expenditures would increase from $71 billion in the goal case to $105 billion in 2066 as more IGCC + CCS plants would have to be built in that year in this goal delayed scenario.

4.9 GHG Relief for Other Sectors

If electricity generation GHGs were reduced to below the target of 784 M tonnes of CO_2-eq./year, then other sectors could reduce their GHG reduction targets. For example, Table 4.4 shows one possible mixture of sector reductions that meets our overall target of an 80 % reduction below 1990 levels. In this case we assume that electricity is reduced by 83.6 % to yield the 306 million tonnes of CO_2-equivalent GHG emissions by 2050. With this reduction in utility emissions, then the industrial sector reduction goal could be cut from the 75 % assumed in Chap. 3 to a reduction of only 50 % below 1990 levels, and the Residential and Commercial sectors could be reduced from a 75 % reduction to only 70 % reductions each, which yields a net overall U.S. societal reduction of 80 %. Of course other combinations are possible if the commercial or residential sectors could not achieve their assumed reductions of 70 % below 1990 levels.

In Appendix A we estimate the probable utility or independent power producer (IPP) return on investment as we make the transition from today's grid mix to a future mix with low carbon generators. We show that coal-powered IGCC + CCS generators with a modest carbon tax could make a 20-year return on investment of more than 11 %, which surpasses the average 2015 rate of 10 % allowed by Public Unity Commissions (PUCs) in those remaining electricity markets still regulated by PUCs [11, 12].

Table 4.4 One possible combination of sector GHG reductions to achieve the societal goal of an 80 % over-all reduction below 1990 levels, assuming that the electricity sector achieved the 307 million tonne/year emission rate calculated above with NGCC + CCS plants replacing some IGCC + CCS generators

	1990 GHGs by sector (electricity separate)	1990 GHG %'s	Suggested % reductions below 1990	Net GHG Goals (M tonnes CO_2)
Residential	345.4	6.4	*70.0*	103.6
Commercial	385.3	7.2	*70.0*	115.6
Industrial	1531.5	28.5	*50.0*	765.8
LDVs	994.0	18.5	*97.1*	28.8
Other on-road transport	239.5	4.5	*92.0*	19.2
Other transport	322.9	6.0	*88.0*	38.7
Electricity	1866.1	34.7	*83.6*	306.0
Agriculture	−313.0	−5.8	*0.0*	(313.0)
Total GHGS	5371.7			1064.7
Net % reduction in total GHG emissions below 1990 levels				−80.2 %

4.10 Conclusions

4.1. Although the coal-powered IGCC plant costs more than a natural gas-powered plant, the incremental costs of adding CCS to a coal-powered IGCC plant per gram of GHG reduction are lower ($1.44/kW per g/kWh saved) than those of an NGCC plant ($1.75/kW per g/kWh saved.)

4.2. The capital costs of replacing all coal, oil and natural gas combustion turbines with a combination of natural gas combined cycle (NGCC) and coal-powered IGCC plants with CCS are large at $77 billion per year in 2070 and 2071 with the parameters used in this simulation, but this is consistent with the US electric utility industry total expenditures of up to $149 billion (including transmission, distribution, environmental and other) in 2014.

4.3. The investor owned utilities and other independent power producers (IPPs) could make an adequate 20-year return on investment of at least 11 % for IGCC + CCS power plants (See Appendix A for details).

References

1. Finkenrath M (2011) Cost and performance of Carbon Dioxide capture for power generation. International Energy Agency, Paris, France 2011. Available at http://www.oecd-ilibrary.org/energy/cost-and-performance-of-carbon-dioxide-capture-from-power-generation_5kgggn8wk05l-en?crawler=true&mimetype=application/pdf
2. Table 8.2 of the annual energy outlook-2015 by the Energy Information Agency of the U.S. Department of Energy. Available at https://www.eia.gov/forecasts/aeo/assumptions/pdf/table_8.2.pdf
3. Bedilion R Program on technology innovation: integrated generation technology options - 2012. Electric Power Research Institute report 1026656, updated February 2013. Available at http://www.epri.com/abstracts/Pages/ProductAbstract.aspx?productId=000000000001026656
4. World Energy Investment Outlook-2014. The International Energy Agency, Paris, France. Available at https://www.iea.org/publications/freepublications/publication/WEIO2014.pdf
5. Klara J (2007) Cost and performance base line for fossil energy plants. DOE/NETL-2007/1281, Rev 1 Aug 2007. Available at https://www.netl.doe.gov/File%20Library/Research/Energy%20Analysis/Coal/BitBase_FinRep_2007.pdf
6. The IHS Power Capital Cost Index (PCCI) for North America. Available at https://www.ihs.com/info/cera/ihsindexes/
7. Wang M et al The greenhouse gases, regional emissions, and energy uses in transportation (GREET) Model, the Argonne National Laboratory. Available at https://greet.es.anl.gov/
8. Overton T (2016) Kemper county IGCC costs rise and delays loom—again. Power Magazine, 5 Apr 2016. Available at http://www.powermag.com/kemper-county-igcc-costs-rise-and-delays-loom-again/
9. Mufson S (2014) Intended showcase of clean-coal future hits snags. The Washington Post, 17 May 2014. Available at https://www.washingtonpost.com/business/economy/intended-showcase-of-clean-coal-future-hits-snags/2014/05/16/fc03e326-cfd2-11e3-b812-0c92213941f4_story.html

10. Rosenberg W IGCC Financing, Chapter 2. Belfer Center at Harvard University. Available at http://belfercenter.hks.harvard.edu/files/igcc%20financing%20chapter%202.pdf
11. Borenstein S, Bushnell J The U.S. electricity industry after 20 years of restructuring, The Energy Institute at HAAS. Available at https://ei.haas.berkeley.edu/research/papers/WP252.pdf
12. What is a wholesale electricity market? Electric Power Supply Association (EPSA). Available at https://www.epsa.org/industry/primer/?fa=wholesaleMarket

Chapter 5
Petroleum and Coal Proven Reserves: The Case for Coal and the Demise of OPEC

Abstract The world now relies almost exclusively on petroleum products to power our transportation systems, and coal, natural gas, nuclear and some petroleum products along with a small but growing contribution from renewables for most electricity production. The petroleum dependence is hazardous since approximately 82.6 % of all petroleum proven reserves reside in the twelve OPEC nations (Algeria, Angola, Ecuador, Iran, Iraq, Kuwait, Libya, Nigeria, Qatar, Saudi Arabia, UAE, Venezuela) based predominantly in the volatile Middle East. (The ten Middle East and North African nations (excluding Ecuador and Venezuela) still account for 72.2% of all proved petroleum reserves). Coal, on the other hand, is widely distributed around the globe, and natural gas is more widely distributed than petroleum. OPEC nations account for at most 3.7 % of proven coal reserves. We conclude that moving to a global energy system based predominantly on coal instead of petroleum would be far more stable for world security.

5.1 Petroleum Reserves

As shown in Fig. 5.1, approximately 83 % of proved oil reserves are in the 12 OPEC nations at the end of 2014 with only 6.7 % in Europe and Eurasia, and 7.9 % in North America including 1.7 % in the U.S. [1]. Note that the OPEC nations have been subtracted from the regional totals for Fig. 5.1; if we add back the OPEC nations into their respective regions, then the Middle East total is 27.7 %, South America total is 11.3 % and Africa is 4.4 %.

5.2 Coal Reserves

Coal reserves are much more uniformly distributed around the globe as shown in Fig. 5.2.

C.E. Thomas, *Stopping Climate Change: the Case for Hydrogen and Coal*,
Lecture Notes in Energy 35, DOI 10.1007/978-3-319-31655-0_5

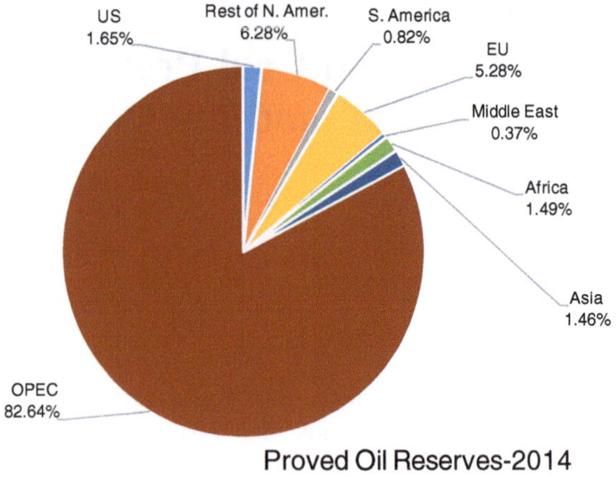

Fig. 5.1 Distribution of global proved oil reserves at the end of 2014

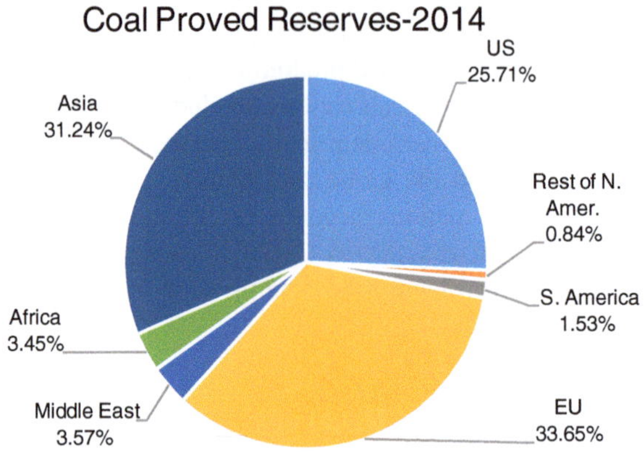

Fig. 5.2 Proved global coal reserves at end of 2014

U.S. proved reserves jump from only 1.7 % of proved oil to 26.7 % of proved global coal reserves.

5.3 OPEC Coal Versus Petroleum

The BP statistical review [1] does not list the coal reserves for all OPEC nations, so we cannot estimate the coal reserves for all OPEC nations, but the entire Middle East and Africa accounts for only 7 % of all coal proved reserves. OPEC nations would clearly lose the dominance in fossil fuels if the world shifts from a primarily oil-based economy to a primarily coal-based economy for both electricity generation and transportation with hydrogen made from coal, nuclear and renewables.

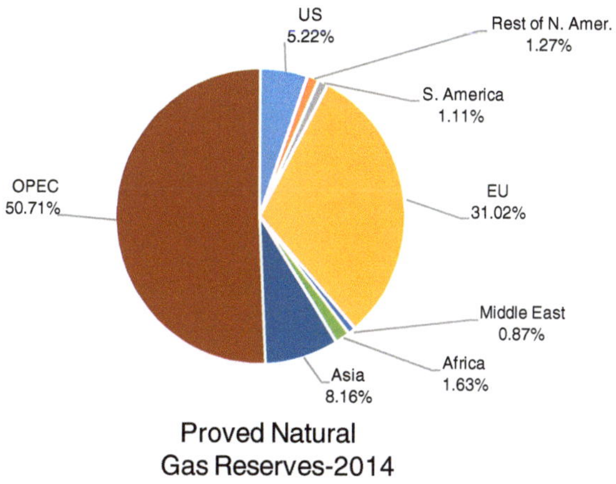

Fig. 5.3 Global distribution of natural gas proved reserves at end of 2014

While we do not have detailed data on coal reserves for all OPEC nations,[1] it is clear that they would no longer control fossil fuel trade, with their 62.1 % share of petroleum reserves to at best 7 % of coal reserves for the entire Middle East and Africa.

5.4 Natural Gas Reserves

Natural gas proved reserves are somewhat better distributed around the globe than petroleum, although more than 50 % of all natural gas reserves are still found in the OPEC nations as shown in Fig. 5.3, much less than the 83 % of all petroleum held by OPEC. The US percentage of natural gas is also slightly higher at 5.2 %, compared to only 1.7 % of oil reserves.

5.5 Ratios of Natural Gas and Coal Percentage Reserves to Petroleum Percentage Reserves

The ratios of percentage of natural gas and coal reserves to percentage of petroleum reserves in each region are summarized in Table 5.1. Thus the US has 3.2 times larger natural gas percentage reserves than petroleum percentage of global petroleum reserves, and 16 times larger percentage of global coal reserves than percentage of global oil reserves.

[1]OPEC would essentially become irrelevant in a coal/hydrogen society.

Table 5.1 Ratios of percentage global natural gas reserves and percentage of global coal reserves to percentage of global oil reserves

	NG	Coal
US	3.2	15.5
EU	5.9	6.4
Asia	5.6	21.4

Table 5.2 Remaining years of fossil fuels assuming fixed consumption at 2014 consumption rates, and no fuel imports for any region

	Oil	NG	Coal
US	7.0	12.9	523.4
Total N. America	34.4	12.8	501.3
S. & C. America	6.4	6.9	463.8
Europe & Eurasia	24.3	72.6	651.7
Middle East	279.2	171.6	115.5
Africa	95.4	117.8	322.8
Asia Pacific	4.0	22.5	103.8
Average	64.4	59.6	383.2

5.6 Years of Remaining Fossil Fuels at Current Consumption Rates

The number of years of remaining fossil fuels at 2013 consumption rates are shown in Table 5.2 for each region.[2] Thus the U.S. would consume all of its proved oil reserves in 7 years at 2014 consumption rates without any imports, and all of its natural gas reserves in 12.9 years, while proved U.S. coal reserves would last for more than 500 years at 2014 consumption rates. On the average, the world has 64 years of oil remaining at 2014 consumption rates, 60 years of natural gas remaining and over 380 years of coal remaining at 2014 consumption rates.

5.7 Fossil Fuel Reserves with GHG Reductions

The times to exhaustion of natural gas and coal reserves listed above assume the then-current 2014 consumption rates and *proved* reserves, not total reserves. But to reduce GHGs according to Chap. 4, we would have to *increase* coal consumption in IGCC + CCS plants over 2014 rates to achieve our goal of an 80 % reduction below 1990 levels, and we would have to increase both coal and natural gas consumption in the minimum GHG case. This would decrease the time until we exhaust known fossil fuel reserves. In Chap. 4, we analyzed two cases: a "goal GHG" case that just met

[2]These calculations assume no imports to any region, and do not account for future fossil fuel discoveries or changes in fuel consumption rates.

Table 5.3 Reserves of coal (million tons) and natural gas (Trillion cubic feet) and year of depletion under the two GHG reduction goals

	Reserves (Tcf)	Year of depletion	
		GHG goal case	Min GHG case
Proved NG reserves	388.8	2031	2031
NG TRR	2277	2104	2089
	M tons		
53 % of coal reserves	253,340	2182	2258

TRR Total recoverable reserves

our goal of cutting GHG's from generating electricity below 784 million tonnes of CO_2 emissions by 2050, and a "Min GHG" case which would cut electricity GHGs even further to 307 million tonnes by 2050, which would allow other sectors of the economy to reduce their required GHG reduction targets.[3] The natural gas reserves would therefore run out sooner under this "Min GHG" scenario as illustrated in Table 5.3, since NGCC + CCS plants take the place of some IGCC + CCS plants. The proved reserves of natural gas would run out by 2031 in either GHG case. However, the totally recoverable reserves (TTR) of natural gas would last until 2104 in the "GHG goal" case and until 2089 in the "Min GHG" case [2].

In the case of coal, the total known reserves or "demonstrated reserve base (DRB) are estimated at 478,000 million tons, but

> Because of property rights, land use conflicts, and physical and environmental restrictions,

EIA has estimated that only about 53 % of the DRB may be "available or accessible for mining" [3] or a total of 253,340 million tons, which would last until 2182 for the Goal case and until 2258 in the "Min GHG" case where NGCC + CCS plants replace some IGCC + CCS plants, thereby reducing the use of coal. We have assumed that coal is also consumed in making hydrogen each year for up to 253 million FCEVs in the U.S. for both cases, which adds 410 million tons of coal consumed each year by 2100. In all cases, we assume that the fuel consumption rates after 2100 continue at 2100 levels into the future. And both calculations assume a continuation of the non-electricity consumption of coal and natural gas projected by the EIA in their 2015 Annual Energy Outlook to 2040

[3]In this "Min GHG" case, the residential and commercial sectors' GHG reduction goals could be reduced from 75 to 70 %, while the Industrial sector GHG reduction goal could be reduced from 75 to 60 %; other combinations are possible; the "Min GHG" case provides more flexibility in meeting our GHG goals.

5.8 Conclusions

Moving toward more use of coal and natural gas instead of petroleum for society's energy needs would:

- Improve world security by nearly eliminating dependence of fossil fuels from the volatile Middle East.
- Extend the number of years of remaining *proved* fossil fuels reserves from an average of 76 years for petroleum to more than 50 years for natural gas and more than 400 years for coal at current consumption rates giving society ample time to develop affordable zero-carbon alternatives[4] (see next bullet for remaining years under GHG goal scenarios).
- Allow the U.S. to meet the GHG goal of cutting GHGs by 80 % below 1990 levels by 2050, with enough natural gas to last through 2104 for the "Goal GHG" case and through 2089 for the "Min GHG" Case, and enough domestic coal to last through 2182 and 2258 for those two cases.

References

1. BP Statistical Review of World Energy, June 2015. Available at https://www.bp.com/content/dam/bp/pdf/energy-economics/statistical-review-2015/bp-statistical-review-of-world-energy-2015-full-report.pdf
2. EIA Table 9-2 Technically recoverable U.S. dry natural gas reserves as of Jan 1, 2013. Available at http://www.eia.gov/forecasts/aeo/assumptions/pdf/oilgas.pdf
3. U.S. Coal reserves, The U.S. Energy Information Administration, Mar 24 2015. Available at http://www.eia.gov/coal/reserves/

[4]Such as nuclear fusion and renewables to produce hydrogen by electrolysis of water, with hydrogen storage to permit increased use of intermittent renewables; the stored hydrogen would produce electricity through a stationary fuel cell system when the sun was not shining and the wind was not blowing, or if renewable energy was not sufficient to supply the full electrical load.

Chapter 6
Greenhouse Gas Reduction Costs Compared: Electricity Versus Fuel Cell Electric Vehicles

Abstract We must cut GHGs in *both* transportation and in the generation of electricity to achieve our GHG reduction goal of 80 % below 1990 levels. In this chapter we explore the costs of cutting GHGs by installing cleaner electricity generation plants and we also present the costs of cutting GHGs in the transportation sector by installing hydrogen infrastructure to power zero-emission fuel cell electric vehicles.

6.1 Cost of Reducing Electric Power Generation Greenhouse Gases

6.1.1 Total Cost of Installing Electricity Generators

We use an average of five sources from the literature[1] to estimate the cost in $/kW for the main electricity generators considered here, as shown in Table 6.1

If the U.S. were to shut down all existing fossil fuel-powered electricity generators[2] (712 GW nameplate capacity in 2015)[3] [1] and replace them with other generators, the costs would be excessive as summarized in Table 6.2, assuming that each generator type replaced *all* 712 GW fossil fuel capacity. The GHG emissions listed in the last column of Table 6.2 are taken from the Argonne National Laboratory's "well-to-wheels" GREET model [2]. For near-zero carbon nuclear power, the total cost would be almost $4 trillion, or $79 billion per year for 50 years. For the next lowest carbon option, NGCC plus CCS at 50.9 g CO_2-equivalent per kWh, the total costs would be $1.4 trillion. Spread over 50 years, the annual cost would be $28 billion, which is much less than current US electric utility spending on generators as shown in the last column of Table 6.3, which averaged just over $53 billion per year from 2012 to 2015. If we use a

[1]See Chap. 4 after Table 4.3 for references to the five sources for electricity generator costs.
[2]Excluding distributed generation by individuals and companies that would not be affected by electric utilities.
[3]Total Nameplate Capacity in 2015 was 1009 GW including renewables and nuclear power.

C.E. Thomas, *Stopping Climate Change: the Case for Hydrogen and Coal*,
Lecture Notes in Energy 35, DOI 10.1007/978-3-319-31655-0_6

Table 6.1 Estimated electricity generator costs in \$/kW from the literature, all converted to 2015 dollars/kW

	IEA-2011	AEO-2015	EPRI-2015	WEIO-2020	NETL-2007	Averages
Coal-ST	\$2285	\$2917	\$2433	\$2123	\$1641	\$2280
IGCC	\$2733	\$3747	\$3287	\$2628	\$1934	\$2866
IGCC + CCS	\$3925	\$6527	\$4152	\$3811	\$2622	\$4207
NGCC	\$1014	\$1022	\$1084	\$1314	\$582	\$1003
NGCC + CCS	\$1812	\$2083	\$2110	\$1820	\$1231	\$1811
Nuclear	ND	5413.5	\$5791	\$5054	ND	\$5420

Table 6.2 Estimated annual expenditures to replace all US fossil fuel utility electric generators over 50 years

Generator Type:	\$/kW	Total costs (\$ Trillion)	Cost per year spread over 50 years (\$B)	GHG (g CO_2/kWh)
IGCC	\$2866	\$2.0	\$40.8	1112.1
IGCC + CCS	\$4207	\$3.0	\$59.9	87.4
NGCC	\$1003	\$0.7	\$14.3	507.2
NGCC + CCS	\$1811	\$1.29	\$25.8	50.9
Nuclear	\$5420	\$3.86	\$77.2	14.0

combination of the two options with carbon capture and sequestration (CCS), the average annual cost would be \$46.4 billion, or slightly less than the average utility expenditures from 2012 to 2015.

The first column of Table 6.3 shows all estimated capital expenditures [3] for investor-owned utilities (IOUs) only.[4] The second column shows the inferred capital expenditures for the entire utility industry based on the estimate that IOUs accounted for 62 % of all electric utility expenditures[5] as shown in Table 6.4 [4]. The fourth column of Table 6.3 shows estimated capital expenditures for generators only, estimated at 37 % of all expenditures, based on the Edison Electric Institute estimate of the split of electricity utility expenditures between generation, distribution, transmission, natural gas related expenditures and "other" [4] as summarized in Table 6.5

In reality, many power plants will reach the end of their useful lives in the coming few decades, and would have to be replaced even without any emissions criteria, so incremental generator costs will be *less* than shown here.

Cost of reducing motor vehicle greenhouse gases with hydrogen fuel cell electric vehicles

[4]The Edison Electric Institute estimated the 2014 and 2015 capital expenditures at approximately \$93 billion and \$85 billion for the IOUs only [4].

[5]The IOUs account for 36.2 % of all U.S. electricity capital expenditures, but 41.7 % of this nameplate capacity resides with non-utility or privately owned generators that will not be affected by changing generation equipment in the utility sector.

Table 6.3 Estimated annual capital expenditures on electrical generators by the electric utilities (last column) in US$ Billions (third column includes distribution, transmission and other expenditures) IOU = Investor-owned utilities

	IOU's only	Inferred total industry	Inferred generator expenditures
	$B	$B	$B
2004	41.1	66.1	24.5
2005	48.4	77.9	28.8
2006	59.9	96.4	35.7
2007	74.1	119.2	44.1
2008	82.8	133.2	49.3
2009	77.6	124.9	46.2
2010	74.2	119.4	44.2
2011	78.6	126.5	46.8
2012	90.3	145.3	53.8
2013	90.3	145.3	53.8
2014	93	149.6	55.4
2015	85	136.8	50.6

Table 6.4 Estimated nameplate capacities of different electricity generating entities

	All generators		Utilities only	
	(GW)	%	(GW)	%
Publicly owned utilities	120.3	10.3	120.3	17.7
Investor-owned utilities	423.1	36.2	423.1	62.2
Cooperatives	61.4	5.3	61.4	9.0
Federal power agencies	75.9	6.5	75.9	11.2
Non-utility generators	487.9	41.7		
Totals:	1168.7		680.8	

Table 6.5 Split of IOU expenditures between different functions

Generation	37 %
Distribution	21 %
Transmission	17 %
NG-related	12 %
Environmental	7 %
Other	6 %

There are two costs associated with making a transition from the current light duty vehicle fleet of mostly gasoline-powered vehicles to the only option [5] that could reasonably be expected to achieve an 80 % reduction in GHGs in the transportation sector in the next few decades: the hydrogen-powered fuel cell electric vehicle (FCEV).[6] One cost to make the transition to FCEVs is the cost of installing a large-scale hydrogen infrastructure. The second cost is the incremental cost of the FCEV compared to a conventional gasoline internal combustion engine vehicle.[7]

6.1.2 Hydrogen Infrastructure Cost Estimates

In a previous book [5], we estimated the costs of building a distributed hydrogen fueling infrastructure system similar to our existing gasoline and diesel fuel infrastructure. The least costly hydrogen fueling station (trucked-in liquid hydrogen in the U.S.) cost an average of approximately $1059 per vehicle.

A very detailed analysis by McKinsey and Company [6] estimated that the cost of installing hydrogen infrastructure in the European Union would be €1010 per FCEV or approximately $1131/FCEV.[8] Therefore, to provide hydrogen infrastructure to support the replacement of all 253 million U.S. light duty vehicles with FCEVs would require expenditures of between $268 billion to $286 billion or an average of $277 billion. Spread over 40 years, this would amount to an average of $6.9 billion per year for hydrogen infrastructure, which is only 13 % of the annual utility industry capital expenditures of $53.4 billion per year from 2012 to 2015 for electric generators.

For comparison, installing battery charging infrastructure to charge battery electric vehicles (BEV) would cost five times as much as installing a hydrogen infrastructure according to the McKinsey & Company report [6]. They estimated that it would cost €101 billion over 40 years to install a hydrogen infrastructure for Europe, while it would cost €540 billion to install a BEV charging network to support BEVs in Europe.

6.1.3 Fuel Cell Electric Vehicle Incremental Cost Estimates

Initially FCEVs will cost more than conventional gasoline or diesel internal combustion engine vehicles. For example, Kromer and Heywood at MIT have estimated

[6]As shown in our previous book on transportation options [5] battery electric vehicles (BEVs) could not achieve an 80 % GHG reduction due primarily to low BEV market penetration capability, as a result of limited range, long refueling times, and limited assigned parking places where batteries could be charged overnight.

[7]This incremental vehicle cost would be paid by car owners who decide that FCEVs offer an advantage over other vehicles, just as they now pay a premium for hybrid electric vehicles with high fuel economy; nonetheless, we treat the incremental cost of FCEVs as a societal cost to be included with the cost of building the hydrogen infrastructure.

[8]Assuming the current exchange rate of one Euro = US$1.12.

Table 6.6 Estimated vehicle Total Costs of Ownership (TCO) by McKinsey & Company (*top half*) and average costs weighted by estimated US market shares of the vehicle segments (*lower half*) including conversion from Euros to dollars

Segment:	A/B	C/D	J			
FCEV	€21,400	€32,800	€41,400			
Gasoline ICV	€18,800	€32,000	€43,200			
Diesel ICV	€18,900	€32,900	€44,600	1€ =	$1.12	FCEV
Market share	25.3 %	28.2 %	46.5 %	Wgt'd ave €	Wgt'd ave $	Added cost
FCEV	€5416	€9252	€19,244	€33,912.3	$37,981.8	
gasoline ICV	€4758	€9026	€20,081	€33,865.3	$37,929.2	$52.63
Diesel ICV	€4783	€9280	€20,732	€34,795.3	$38,970.7	$(988.9)

that a FCEV with 390 miles range would cost approximately $3600 more than a conventional vehicle [7].[9] Thus to replace all 253 million U.S. vehicles with FCEVs would cost an additional $911 billion. Spread over 40 years, this would amount to approximately $22.8 billion per year. Hence the total of hydrogen infrastructure costs and incremental FCEV costs would average to approximately $29.7 billion per year, which is only 43 % of the average utility industry expenditures of $53.4 billion per year on electrical generators from 2012 to 2015.

The McKinsey & Company report is particularly credible on the cost of FCEVs, since 11 auto companies[10] provided McKinsey with confidential vehicle cost data in "clean rooms" to avoid giving their competitors proprietary cost data [6]. They concluded that by 2050 the total cost of ownership (TCO) for large "J-segment" FCEVs would cost *less* than conventional cars (€41,400 for FCEVs vs. €43,200 for gasoline ICVs and €44,600 for diesel ICVs. For A/B segment[11] cars, they estimated that the TCO for FCEVs would be slightly more expensive than gasoline or diesel internal combustion engine cars as summarized in the top half of Table 6.6, while, for C/D segment[12] cars the FCEV TCO would be €800 more than a gasoline ICV but €100 less than a diesel ICV.

The US light duty vehicle fleet is dominated by large vehicles similar to the J-class, including SUVs, vans and even pickup trucks. The latest U.S. data on vehicle sales by size class[13] is from 2012, where the Oak Ridge National Laboratory reported sales of all small vehicles at 25.3 %, medium and large

[9]Kromer and Heywood also estimated that a battery electric vehicle with only 200 miles range would cost approximately $10,200 more than a conventional vehicle [7].

[10]BMW, Daimler, Ford, GM, Honda, Hyundai, Kia, Nissan, Renault, Toyota, and Volkswagen.

[11]Similar to sub-compact cars in North America.

[12]Similar to compact and mid-size cars in North America.

[13]The Oak Ridge data book relies on data from the EPA, and the EPA no longer reports vehicle sales and vehicles on the road by class size.

Table 6.7 Estimated cost of GHG reductions by replacing 253 million gasoline hybrids with fuel cell electric vehicles

	Hydrogen Infrastruct. Cost ($B/yr)	Incremental FCEV cost	Total cost	GHG Savings (B tonnes/yr)	GHG Cost savings ($/tonne)
MIT	267.9	6.9	274.8	0.090	$3047.9
McKinsey	286.1	$0.3	286.5	0.090	$3177.1
Averages	277.0	3.6	280.7	0.090	$3112.5

vehicles[14] at 28.2 %, and larger SUVs, pickup trucks and vans at 46.5 % of all U.S. sales [8]. Using this vehicle size distribution, the weighted average vehicle cost estimates from McKinsey & Company are shown on the bottom half of Table 6.6. Converting Euros to US dollars with a $1.12 exchange ratio yields an estimate that, on the average, a FCEV TCO would be only $52.63 more than a gasoline ICV that is typical in the US, and the FCEV TCO would be $989 *less* than a diesel ICV that is more prevalent in the EU.

So converting all 253 million U.S. LDVs to FCEVs would have a TCO of $13.3 billion or only $333 million per year over 40 years according to the McKinsey & Company estimates.

As shown in Table 6.7, the average cost of reducing GHGs with FCEVs is $3,112 per tonne of GHG reductions.

Note that the total cost of converting all vehicles to BEV would include both the incremental cost of the BEV and the cost of "greening the grid," since adding BEVs with today's U.S. grid that is primarily burning fossil fuels would *increase* GHGs on the average [5].

6.2 Summary of Estimated Costs for Reducing GHG Emissions

The total GHG reductions and capital costs for converting the electricity grid to low-carbon generators are summarized in Table 6.8. The total costs of reducing GHGs by installing lower carbon electricity generators is approximately $1350.5/tonne of CO_2-equivalent reductions, which is approximately 43 % of the societal cost of replacing gasoline hybrid vehicles with hydrogen-powered fuel cell electric vehicles[15] at $3112.5/tonne (Table 6.7).

---

[14]For the purposes of the market share percentages, we assumed that US. "large" passenger vehicles were included in the European "C/D segment."

[15]Including both hydrogen infrastructure cost and the incremental cost of FCEVs compared to gasoline hybrid electric vehicles (HEVs).

Table 6.8 Estimated cost of GHG reductions with lower carbon electricity generators (Min GHG Case)

Total GHG reductions (M tonnes/yr)	1940.8
Total cost ($B)	2621.1
GHG cost ($/tonne)	1350.5

6.3 Greenhouse Gas Reductions with Alternative Vehicles

In a previous book, we calculated the GHG reductions possible with various alternative fuels and vehicles [5]. The results for the estimated US electricity grid mix for the 2020 time period are summarized in the second through fourth columns of Table 6.9. We assume that all hydrogen is made from natural gas, the current

Table 6.9 Alternative vehicle GHGs for the 2020 time period (second through fourth columns) and for vehicles with both electricity and hydrogen produced by coal-based IGCC + CCs plants (last three columns)

				H2 and Electricity from		
	H2 from NG			from IGCC + CCS		
	Vehicle wgt'd average GHGs (g/mile)	% GHGs relative to gasoline HEVs	% GHGs relative to gasoline ICVs	Vehicle wgt'd average GHGs (g/mile)	% GHGs relative to gasoline HEVs	% GHGs relative to gasoline ICVs
Gasoline ICV	549.8	62.2	0.0	549.8	62.2	0.0
Gasoline HEV	339.0	0.0	−38.3	339.0	0.0	−38.3
Gasoline PHEV-40	327.7	−3.3	−40.4	237.4	−30.0	−56.8
E-85 ICV	298.6	−11.9	−45.7	298.6	−11.9	−45.7
Diesel ICV	474.3	39.9	−13.7	474.3	39.9	−13.7
Diesel HEV	307.9	−9.2	−44.0	307.9	−9.2	−44.0
NGV	488.0	44.0	−11.2	488.0	44.0	−11.2
NG HEV	309.7	−8.6	−43.7	309.7	−8.6	−43.7
NG PHEV40	332.1	−2.0	−39.6	224.4	−33.8	−59.2
BEV	371.4	9.6	−32.4	87.5	−74.2	−84.1
FCEV	188.2	−44.5	−65.8	3.3	−99.0	−99.4

ICV = Internal Combustion engine Vehicle; E-85 = mixture of 85 % ethanol in gasoline; HEV = hybrid electric vehicle; NGV = natural gas vehicle; NG = Natural Gas; BEV = Battery Electric Vehicle; FCEV = Fuel Cell Electric Vehicle; PHEV-40 = plug-in hybrid electric vehicle with 40 miles all-electric range lowest cost option, and electricity for electric vehicles comes from the *marginal* U.S. grid mix.

The distinction between average and marginal grid mix is important. For example, consider a hypothetical small utility that has only two generators: a hydroelectric generator with zero GHG emissions, and a coal-powered generator with approximately 1000 g of CO_2-equivalent GHG emissions per kWh of electricity generated. Some analysts might use the average of these two generators or 500 g/kWh. However, in reality, to maximize income utilities usually dispatch their generators with the lowest operating cost first; in this case the hydroelectric plant. The coal plant would only be used to supply added capacity if needed. If the utility load was approximately half their total capacity, then the hydroelectric plant would supply most of their demand. But if a new load is added, such as to charge battery electric vehicles (BEVs), then the coal plant would have to be turned on.[16] The *marginal* generator is therefore the coal plant. In effect, adding BEVs will increase the utilities' GHG emissions from the coal plant at 1000 g/kWh, or twice the average GHG emissions of 500 g/kWh.

Calculating the marginal grid mix is very complicated, since the analyst must determine which generator is on the margin during each time period every day. Fortunately, researchers at the Oak Ridge National Laboratory [9] made such a calculation for the various electric grid regions in the U.S. In some regions, GHGs increased for BEVs, while in other regions with less coal power, BEV GHGs decreased relative to a conventional car. We weighted the GHGs in each power region by the number of light duty vehicles in each region. Unfortunately, the regions with the most vehicles had the higher GHG emissions. For example, the old[17] Southeastern Reliability Council region was home to 17 % of all US light duty vehicles; at the time of the Oak Ridge study, fossil fuels were used to generate 98 % of all electricity (65 % natural gas and 33 % coal). As a result, based on the vehicle-weighted average emissions, the average BEV is the US has 9 % higher GHGs than a gasoline hybrid electric vehicle (HEV), 371–339 grams per mile[18]as shown in Table 6.9

Note also from Table 6.9 that plug-in hybrid electric vehicles (PHEVs) generate slightly lower GHGs with lower carbon IGCC + CCS electricity, as would be expected.

[16]In reality the coal plants must always be fired up, since it takes many hours to ramp up a coal plant. So the situation is even worse for this hypothetical utility, since there will be added GHGs to keep the coal plant warmed up even if it is never used to produce electricity.

[17]The Energy Information Administration (EIA) no longer uses these electricity regional designations.

[18]See also Appendix D of reference [5] for more details on the calculation of marginal grid mixes and BEV GHG emissions.

Table 6.10 Estimated GHGs for hydrogen co-production with electricity at three different coal-powered IGCC + CSS plants [9]

	kg CO_2/GJ H2 (LHV)	kg CO_2/kg H2	g CO_2/kg H2	grams CO_2/mile
HPS	12.3	0.11	113.8	1.85
HPS-Y	12	0.11	111.0	1.81
HPS-L	42.1	0.39	389.6	6.34
Average			204.8	3.33

Table 6.11 Measured and average fuel cell electric vehicle fuel economy

Toyota Mirai	66.0	Miles/kg
Toyota Highlander	68.3	Miles/kg
Hyundai Tucson	50.0	Miles/kg
Average:	61.4	Miles/kg

The last three columns of Table 6.9 show the vehicle GHGs when both electricity and hydrogen are made from coal-powered IGCC plants with CCS. The BEV benefits from CCS, with GHGs dropping from 371 to 87.5 grams/mile. But the FCEV benefits even more, assuming that the hydrogen is made at a coal-powered integrated gasification combined cycle (IGCC) plant with CCS that co-produces both electricity and hydrogen. As shown by Chiesa et al. [10] hydrogen produced at an IGCC plant with CCS generates very little GHGs as summarized in Table 6.10; they estimated the GHGs for hydrogen from three different co-generation plants, with an average emission rate of 204.8 g of CO_2-equivalent per kg of hydrogen produced.[19] With an average FCEV fuel economy of 61.4 miles/kg of hydrogen[20] as summarized in Table 6.11, the FCEV would produce only 3.3 g of CO_2-equivalent per mile, or 26 times less than a BEV and almost 102 times less than a gasoline-powered hybrid electric vehicle like the Toyota Prius.

[19]The authors calculated the hydrogen GHGs by assigning most GHGs to the electricity, with an average electricity GHG rating of approximately 73.4 g/kWh of CO_2-equivalent emissions, which is close to the GREET model estimate of 87.4 g/kWh used in our model. The residual GHG emissions were then assigned to hydrogen to arrive at the estimate of 204.8 g CO_2-equivalent per kg of hydrogen produced.

[20]The FCEV fuel economies are taken from the EPA fuel economy estimates [11] for the Mirai and Tucson FCEVs, and from on-road testing by two government laboratories of two Toyota Highlander FCEVs [12].The future fuel economies of FCEVs should be higher than listed here, since (a) two of these vehicles were SUVs, and fuel economy will improve for smaller passenger vehicles due to reduced weight, cross-sectional area, and lower aerodynamic drag, and (b) the measured fuel economy of those two SUV FCEVs were based on decade-old fuel cell technology. With improved fuel cell efficiency and reduced weight, we would expect future FCEV fuel economy to be greater than 61 miles/kg of hydrogen.

References

1. The 2015 Annual Energy Outlook (AEO), Electric power sector, electricity generating capacity. The Energy Information Administration (EIA). Available at: http://www.eia.gov/oiaf/aeo/tablebrowser/#release=AEO2015&subject=6-AEO2015&table=9-AEO2015&region=0-0&cases=ref2015-d021915a
2. Wang M et al (2008) The greenhouse gases, regional emissions, and energy uses in transportation (GREET) model. The Argonne National Laboratory, Available at: https://greet.es.anl.gov/
3. American Public Power Association 2014-15 annual directory and statistical report (based on 2012 data). Available at: http://www.publicpower.org/files/PDFs/USElectricUtilityIndustryStatistics.pdf
4. Edison Electric Institute (2013) Financial review. Available at: http://www.eei.org/resourcesandmedia/industrydataanalysis/industryfinancialanalysis/finreview/Documents/FinancialReview_2013_02_IndustryFinPerf.pdf
5. Thomas CE (2015) Sustainable transportation options for the 21st century and beyond. Springer Publishing book in print, ISBN # 978-3-319-16831-9
6. McKinsey & Company A portfolio of power-train options for Europe: a fact-based analysis: the role of battery electric vehicles, plug-in hybrids, and fuel cell electric vehicles, undated. Available at: http://ec.europa.eu/research/fch/pdf/a_portfolio_of_power_trains_for_europe_a_fact_based__analysis.pdf
7. Kromer M, Heywood J (2007) Electric power train opportunities and challenges for the U.S. light-duty vehicle fleet. Sloan Automotive Laboratory, Laboratory for Energy and Environment, Massachusetts Institute of Technology, Cambridge, Massachusetts, Report # LFEE 2007-03XP, May 2007. Available at: http://web.mit.edu/sloan-auto-lab/research/beforeh2/files/kromer_electric_powertrains.pdf
8. Davis S et al (2013) The transportation energy data book, edition 32. The Oak Ridge National Laboratory, July 2013, Table 4.7. Available at: http://cta.ornl.gov/data/index.shtml
9. Hadley S, Tsvetkova A (2008) Potential impact of PHEVS on regional power generation. Report # ORCL/Tm-2007/150. Jan 2008. Available at: http://www.ornl.gov/info/ornlreview/v41_1_08/regional_phev_analysis.pdf
10. Chiesa P, Consonni S, Kreutz T, Williams R (2005) Co-production of hydrogen, electricity and CO2 from coal with commercially ready technology, Part A: Performance and emissions. Int J Hydrogen Energy 30(7):747–767. Available at: http://www.sciencedirect.com/science/article/pii/S0360319904003052
11. Compare Fuel Cell Vehicles, U.S. Environmental Protection Agency. Available at: https://www.fueleconomy.gov/feg/fcv_sbs.shtml
12. Wipke K et al (2009) Evaluation of range estimates for Toyota FCHV-adv. under open-road driving conditions. Savannah River National Laboratory and the National Renewable Energy Laboratory, Report # SRNS-STI-2009-00446, August 2009. Available at: http://energy.gov/eere/fuelcells/downloads/evaluation-range-estimates-toyota-fchv-adv-under-open-road-driving

Chapter 7
Tri-Generation at Big Box Stores and Warehouses

Abstract Building hydrogen fueling stations for fuel cell electric vehicles (FCEVs) will be challenging initially, before there are enough FCEVs on the road to make these fueling stations profitable. One solution to overcome this "chicken and egg" dilemma is to build distributed generation stations at big box stores or warehouses that supply three fuels: hydrogen, electricity and heat. These "Tri-Gen" stations can be profitable without any FCEVs in the neighborhood primarily by providing electricity and to the stores (heat revenues are minimal), thereby cutting fuel costs to pay for the station. Some stores and warehouses have also converted from battery-powered forklift trucks to fuel cell trucks powered by hydrogen. These forklift trucks consume more hydrogen than FCEVs, providing another source of income to early Tri-Gen stations. Many tens of warehouses already have enough fuel cell lift trucks to make Tri-Gen stations profitable; we do not have to wait for large-scale FCEV sales to begin large scale production of Tri-Gen stations. And even if building Tri-Gen stations at remote warehouses with fuel cell fork lift trucks are not close to clusters of early FCEVs, these warehouse Tri-Gen stations will reduce the cost of station components by increasing production volumes and traveling down the cost "learning curve" as more systems are built, enhancing the economics of Tri-Gen stations in metropolitan areas with more FCEVs. In addition, hydrogen produced profitably at remote warehouses with fuel cell forklift trucks can be shipped to satellite hydrogen fueling stations near large concentrations of FCEVs

7.1 Fuel Cell Fork Lift Trucks

One of the earliest commercially successful applications for hydrogen-powered fuel cells is for material handling equipment (MHE) such as forklift trucks in warehouses and stores. Previously these indoor forklift trucks were powered primarily by batteries. But battery-operated forklift trucks have many disadvantages:

C.E. Thomas, *Stopping Climate Change: the Case for Hydrogen and Coal*,
Lecture Notes in Energy 35, DOI 10.1007/978-3-319-31655-0_7

- The batteries lose power during a shift and do not move products as fast at the end of the shift.
- The very heavy batteries must be removed and recharged after a shift.
- Operators must recharge the batteries, which can take hours to charge and more hours to cool down the batteries after charging in preparation for the next shift.
- Significant space is required in the warehouse to accommodate battery charging and cooling down, thereby reducing warehouse space available for product storage.
- Battery output decreases in cold areas such as freezers.

Hydrogen-powered fuel cell lift trucks overcome all of these disadvantages. Fuel cells do not lose power during operation, can operate in freezers, the hydrogen tanks can be refueled in a few minutes, and the hydrogen dispensers do not require much warehouse space. Instead of large battery rooms, the fuel cell lift trucks require only indoor hydrogen fueling dispensers, similar to gasoline pumps. For these reasons, many warehouse operations have converted from battery to fuel cell lift trucks. Table 7.1 shows a partial list of warehouses in the U.S. that have converted at least part of their fork lift truck fleet to run on hydrogen, along with the number of fuel cell fork lifts per facility. These 36 warehouses have installed almost 6000 fuel cell lift trucks by early 2016. 26 of these facilities have more than 100 fuel cell lift trucks with Walmart having 9 facilities with over 200 lift trucks each, making them potential profitable sites for hydrogen Tri-Gen systems as demonstrated below.

Converting from battery lift trucks to fuel cell lift trucks increases warehouse operating efficiency and reduces costs.

Most of these fuel cell lift trucks currently obtain their hydrogen from trucked-in liquid hydrogen[1] (LH2). The warehouse owner purchasing liquid hydrogen can substantially reduce operating costs by installing an on-site Tri-Gen system to produce electricity and hydrogen from natural gas.

Fuel cell lift trucks consume more hydrogen than fuel cell electric vehicles (FCEVs), thereby increasing the revenue from selling hydrogen. For example, a passenger vehicle is rarely used and is parked most of the time, whereas a fork lift truck may be operated three shifts a day. We estimate that each FCEV consumes approximately 166 kg of hydrogen/year for a FCEV traveling 10,220 miles per year, while on the average, a fuel cell fork lift truck consumes approximately 211 kg/year. In addition, there are already many fuel cell fork lift trucks in operation at various warehouses as listed in Table 7.1, while there are very few regions that have more than 10 FCEVs likely to refuel at one location in the next few years. Therefore a Tri-Gen station placed at a warehouse or store with fuel cell lift trucks will have a higher return on investment than a station relying on a handful of FCEVs as shown below.

[1]Although some fuel cell lift truck owners have installed on-site generated hydrogen by reforming natural gas.

Table 7.1 Partial list of the number of hydrogen-powered forklift trucks (or other fuel cell material handling equipment) deployed in the U.S. as of early 2016

Ace Hardware	65	Wilmer TX
Ace Hardware	71	West Jefferson, OH
Assoc. Wholesale Grocers	297	Kansas City MO
Assoc. Wholesale Grocers	240	Pearl River LO
BMW	275	Greer, SC
Central Grocers	200	Joliet, IL
Coca Cola	81	San Leandro CA
Coca Cola	50	Charlotte, NC
CVS	25	N. Smithfield, RI
Fedex	40	Springfield MO
HEB Grocers	14	San Antonio TX
Kroger	179	Compton, CA
Mercedes Benz	218	Vawe, AL
Proctor & Gamble	148	Greensboro NC
Proctor & Gamble	140	Mehoophany, PA
Sysco	198	Plympton MA
Sysco	110	Riverside, CA
Sysco	129	New Braunfills, TX
Sysco	105	Riverside, CA
Sysco	110	San Antonio, TX
Sysco	60	Central Islips, NY
Sysco	113	Houston, TX
Sysco	129	Phillidelphia, PA
Sysco	102	Front Royal, VA
Walmart	263	Johnston, NY
Walmart	254	Ohio
Walmart	1738	7 locations
Wegmans	278	Pottsville PA
Wholefoods	61	Landover, MD
Winco Foods	163	Modesto, CA
Totals	5856	
Average	162.7	
Max	297	
Min	14	

7.2 Tri-Gen System Description

A block diagram of the Tri-Gen station is shown in Fig. 7.1. A steam methane reformer (SMR) converts natural gas to hydrogen, and a Pressure swing adsorption (PSA) system cleans up the hydrogen to purity sufficient for fuel cells. Some of this hydrogen goes to a stationary fuel cell system that generates electricity for the facility, reducing the expenses for purchasing electricity. Some of the hydrogen is

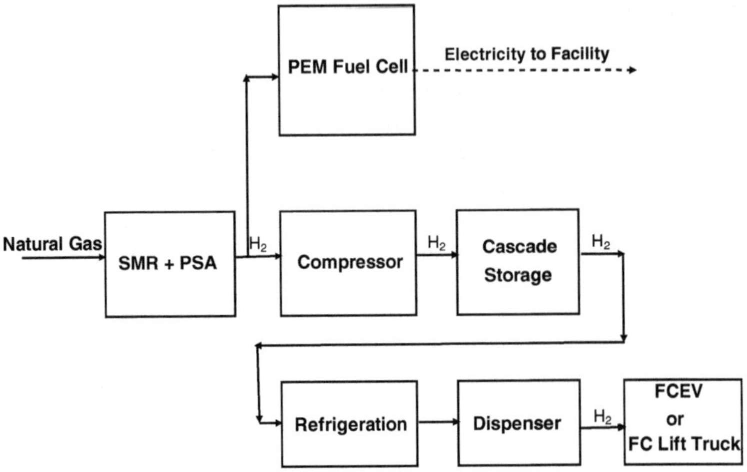

Fig. 7.1 Block diagram of a tri-generation system to support fuel cell lift trucks and FCEVs (heat supplied to the building is not shown here)

Table 7.2 Estimated capital cost for a tri-gen station suitable to provide 330 kW of electricity and to supply enough hydrogen to support up to 100 fuel cell electric vehicles

PEM FC system	$683,332
SMR + PSA	$212,184
Storage	$303,223
Refrigeration	$74,280
Compressor	$475,521
Electrical	$75,233
Dispenser	$68,440
CSD controls and "other"	$286,614
Total Capex	$2,178,827

compressed and stored and later used to fill the hydrogen tanks in fuel cell MHE.[2] In addition, some of this hydrogen can be used for fuel cell electric vehicles (FCEVs) to help start the conversion from gasoline-powered cars to hydrogen-powered cars.

The estimated capital costs for such a tri-gen system from the literature are summarized in Table 7.2; this system is sized to provide 330 kW of electricity, enough to provide all the electricity for an average "Big Box" store or warehouse using fuel cell forklift trucks, plus enough hydrogen to support up to 100 fuel cell electric vehicles.

[2]For material handling equipment (MHE) such as fork lift trucks, the refrigeration is not needed, and the MHE trucks store hydrogen at 300-bar, lower than the 700-bar currently used on FCEVs, so the compressor can be lower power for MHE.

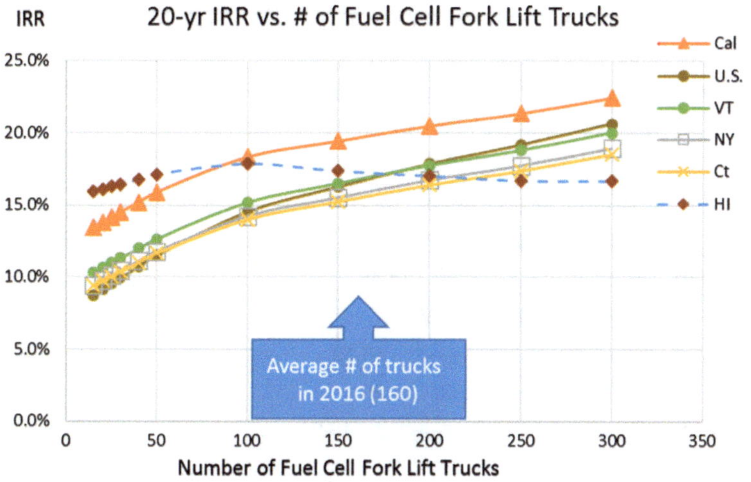

Fig. 7.2 Estimated 20-year IRR for tri-gen stations at warehouses or stores with Fuel cell fork lift trucks (no hydrogen is sold to FCEVs)

7.3 Results and Discussion

The results of this analysis for fuel cell lift trucks are quite promising. The cost of a Tri-gen station based on a stationary low-temperature Proton Exchange Membrane (PEM) fuel cell system to generate electricity for the site is estimated at $2.2 million initially ($668K for the PEM fuel cell system and $1.5 million for the hydrogen station) to provide 330 kWe of electricity for a 105,000 square foot warehouse or store,[3] plus extra hydrogen production capacity to support 100 Forklift trucks using the station. The estimated 20-year internal rate of return (IRR) on this $2.2 million investment is summarized in Fig. 7.2 as a function of the number of fuel cell lift trucks supported by the station. The IRRs depend heavily on the region of the country due to variable electricity and natural gas prices. In general, regions with high electricity rates and low natural gas rates will have the highest returns, since the Tri-Gen station essentially displaces purchased electricity with hydrogen-generated electricity from natural gas. Hawaii has the highest commercial electricity rates in the U.S. as shown in the next section, which yields the highest returns above 15 % even with less than 50 fuel cell lift trucks since the fuel cell distributed generation electricity in Hawaii displaces the most expensive electricity. Many states have IRRs above 15 % with the average number of fork lift trucks at current warehouses in 2016 (160), with California above 19 % as long as there are at least 150 fuel cell lift trucks on site.

Hawaii (dashed line in Fig. 7.2) seems to be an anomaly compared to the other states in Fig. 7.2, with a slower growth and even decreasing IRRs with increased

[3]The average Walmart store is 105,000 square feet.

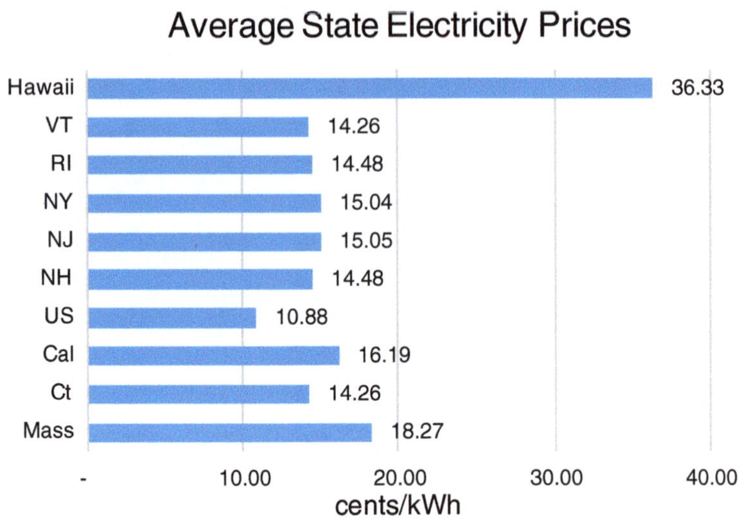

Fig. 7.3 Average State Commercial Electricity Prices in 2014

numbers of lift trucks. This reduced slope is due to the need for a larger and more expensive steam methane reformer (SMR) to support more than 100 fork lift trucks. The impact of the added SMR costs is greatest for Hawaii, but the net effect is a slower increase in IRR as more than 100 lift trucks deployed for all the states as shown in Fig. 7.2.

7.4 State Electricity Rates

The rates for commercial electricity vary widely across the US according to the Department of Energy's Electricity Table Browser [1]. As shown in Fig. 7.3, commercial electricity rates are higher than average in most of the states analyzed in this chapter.

7.5 State Natural Gas Prices

According to Fig. 7.3 Hawaii should have the highest IRR in terms of displaced electricity costs. However, Hawaii also has the highest industrial natural gas prices,[4] as shown in Fig. 7.4 from the Department of Energy's Natural Gas price table

[4]We use industrial natural gas prices since the Tri-Gen system will consume enough natural gas to qualify for lower rates; for example, Washington Gas in Virginia offers reduced rates of 25.8 cents/therm for consumption above 125 therms and 19.7 cents/therm for consumption above

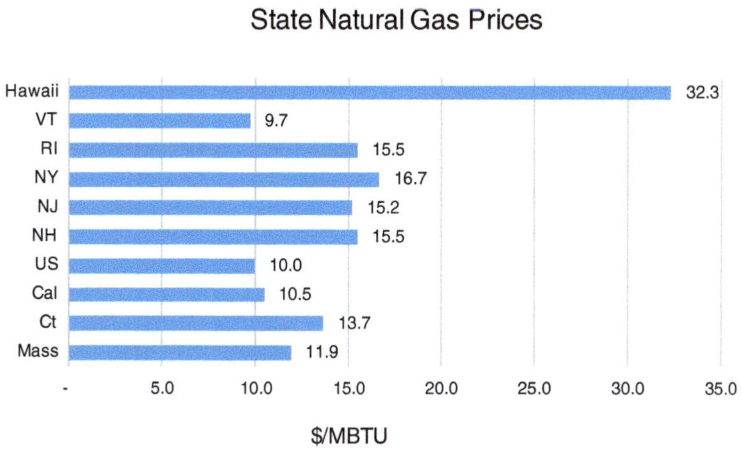

Fig. 7.4 Average State industrial natural gas prices (2015 to 2034)

averaged over 2015 to 2034 (the period used to calculate the 20-year IRRs in Fig. 7.2) [2]. The higher cost of natural gas in Hawaii increases the costs of running the Tri-Gen station and reduces the IRRs. But Hawaii still has the highest returns for small numbers of lift trucks (<100) as shown in Fig. 7.2.

One figure-of-merit often used to compare electricity prices is the "spark spread" or difference between the cost of electricity and natural gas. Since we are using natural gas to make electricity, a large spark spread should indicate higher IRRs for the Tri-Gen station. As shown in Fig. 7.5, the "spark spread" is highest in Hawaii, which confirms its high IRR for small quantities of lift trucks.

7.6 State Gasoline Prices

The price of gasoline is also important for refueling fuel cell electric vehicles (FCEVs) since it sets the price of hydrogen that the station sells.[5] Thus we assume that the price of hydrogen is set such that the driver of a FCEV pays the same cost *per mile* for hydrogen as the driver of a hybrid electric vehicle (HEV) pays for

(Footnote 4 continued)

875 therms, compared to 31.6 cents/therm for small quantities of consumption. For a Tri-Gen station consuming above 17,000 therms/month the natural gas charge is approximately $6.89 per MBTU, which is much closer to the EIA "industrial rate" of $6.43/MBTU than to the "commercial rate" of $9.17/MBTU for Virginia.

[5]Gasoline prices will not affect the rates of return for MHE such as those shown in Fig. 7.1.

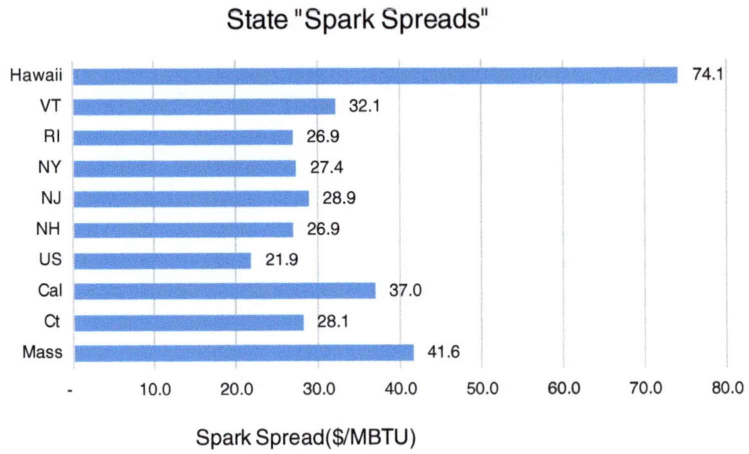

Fig. 7.5 "Spark spreads" or difference between the cost of electricity and the cost of natural gas (Both in $/MBTU)

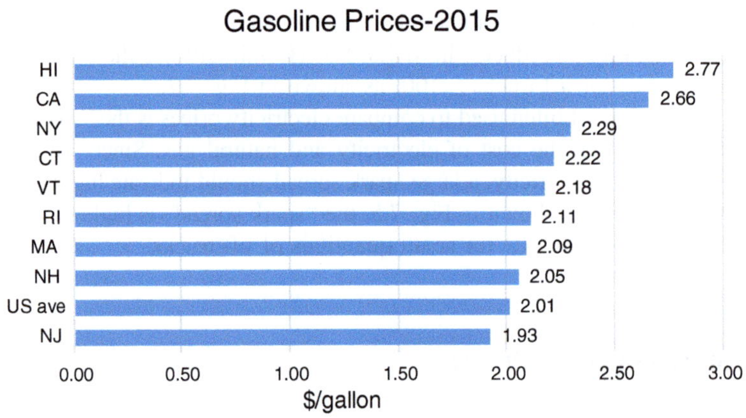

Fig. 7.6 Average state retail prices for regular gasoline including all taxes in 2015

gasoline[6] per mile driven. So hydrogen prices are proportional to gasoline prices. The average prices for regular gasoline in each state in 2015 are shown in Fig. 7.6, including local, state and federal taxes.[7] All of the states considered here have gasoline prices higher than the US average of $2.01/gallon except New Jersey. Hawaii with the highest gasoline price of $2.77/gallon will allow the highest price for hydrogen sold to FCEV owners.

[6]We use gasoline-powered hybrid electric vehicles (HEVs) as the reference gasoline car since automakers will have to sell primarily HEVs to meet the new US fuel economy standards.

[7]We assume that hydrogen would be exempted from taxes initially as an ultra-clean fuel.

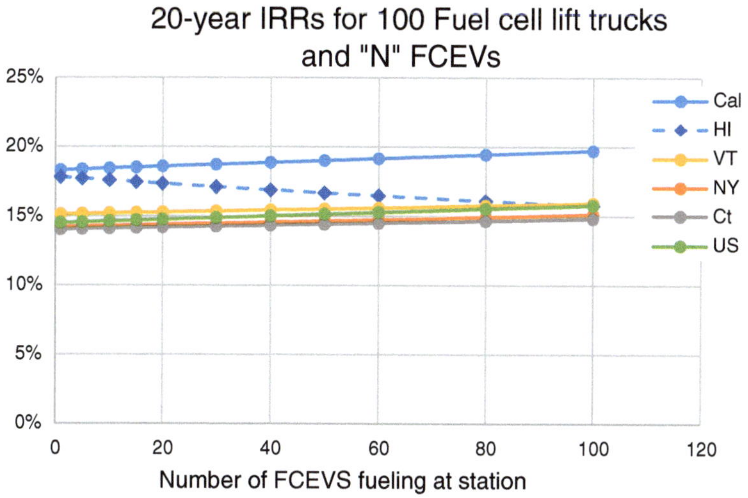

Fig. 7.7 20-year return on investment assuming 100 fuel cell lift trucks plus the number of FCEVs shown consuming hydrogen from the station

7.7 Rates of Return with Fork Lift Trucks and Fuel Cell Electric Vehicles

The previous sections assume that all hydrogen is consumed by fuel cell fork lift trucks. But some warehouses or "Big Box" stores could install hydrogen pumps in their parking lots to provide hydrogen for fuel cell electric vehicle (FCEV) owners shopping at the stores or for fuel cell delivery trucks.[8] The resulting 20-year IRRs as a function of the number of FCEVs using the station with 100 fuel cell lift trucks[9] is shown in Fig. 7.7. In the five states shown in Fig. 7.7, the return on investment exceeds 14 % as long as there are at least 100 Fuel cell lift trucks refueling at the site, and IRRs exceed 17 % in California and Hawaii with 100 or more FCEVs using the station. The 20-year returns in California approach 20 % with 100 FCEVs fueling at the site.

[8]FedEx is developing fuel cell powered delivery trucks to increase the range of their battery-operated delivery trucks.

[9]From Table 7.1, there are already at least 26 warehouses or distribution centers with more than 100 fuel cell lift trucks in operation.

Table 7.3 Electricity and natural gas consumption rates for retail buildings across the U.S

		Miami	Houston	Phoenix	Atlanta	LA	Baltimore	Las Vegas	Seattle	Chicago	Denver	Minneapolis	Helena	Duluth	Averages
equipment	kbtu/ft^2 yr	9.6	9.6	9.6	9.6	9.6	9.6	9.6	9.6	9.6	9.6	9.6	9.6	9.6	9.60
Interior lighting	kbtu/ft^2 yr	48.8	48.8	48.8	48.8	48.8	48.8	48.8	48.8	48.8	48.8	48.8	48.8	48.8	48.80
Exterior lighting	kbtu/ft^2 yr	5.6	5.6	5.6	5.6	5.6	5.6	5.6	5.6	5.6	5.6	5.6	5.6	5.6	5.60
Heating	kbtu/ft^2 yr		0.3		0.1		0.1		0.1	0.1		0.1	0.1	0.1	0.13
cooling	kbtu/ft^2 yr	32.2	24.6	20	13.2	7.4	10.4	8.6	2.5	7.5	6.1	6.3	3.8	2.6	11.17
fans	kbtu/ft^2 yr	19.5	20.2	21.5	18.4	16.8	17.7	20.3	15.7	17.3	19.5	17.2	18.3	16.3	18.36
SWH	kbtu/ft^2 yr	0.2	0.2	21.5	0.3	0.3	0.3	0.3	0.3	0.3	0.3	0.3	0.3	0.4	1.92
Total electric	kbtu/ft^2 yr	115.9	109.3	127	96	88.5	92.5	93.2	82.6	89.2	89.9	87.9	86.5	83.4	95.53
Total electric	kWh/ft^2-yr	33.97	32.03	37.22	28.14	25.94	27.11	27.32	24.21	26.14	26.35	25.76	25.35	24.44	28
Annual electricity	MWh/yr	3281	3094	3596	2718	2506	2619	2639	2339	2525	2545	2489	2449	2361	2705
Gas heating	kbtu/ft^2 yr	0	1.2	0.6	1.8	0.1	4.5	0.9	2.8	7.6	4.5	13	8.7	17.2	
NG use	MBTUs/ft^2-yr	0	0.0012	0.0006	0.0018	0.0001	0.0045	0.0009	0.0028	0.0076	0.0045	0.013	0.0087	0.0172	
Annual NG use	MBTU/yr	–	115.92	57.96	173.88	9.66	434.70	86.94	270.48	734.16	434.70	1,255.80	840.42	1661.52	368
												Average Power required:			335.6 kW

7.8 Rates of Return with Fuel Cell Delivery Trucks

Another option for stores and warehouses is to utilize fuel cell delivery trucks. FedEx is developing fuel cell "range extenders" for electric delivery vehicles [3]. The hydrogen-powered fuel cell is expected to double the 80-mile range of their current battery-powered electric delivery trucks. Since delivery trucks are usually driven more miles than passenger vehicles, fuel cell delivery trucks could consume more hydrogen than FCEVs, which will increase IRRs over those shown in Fig. 7.7. Delivery companies may also find that they can increase the efficiency of trucks with range-extenders to supplement batteries by replacing them with pure fuel cell power. Removing the batteries will reduce the mass of the trucks and increase their fuel economy and reduce the time to recharge the batteries. Fuel cell delivery trucks, like fuel cell fork lift trucks have the added advantage of being centrally located at the warehouse or store, eliminating the uncertainty of having enough FCEVs in their neighborhood to make the venture profitable. The economics for fuel cell delivery trucks should be better than shown in Fig. 7.7 since these delivery trucks will consume more hydrogen than fuel cell passenger vehicles, which are parked most of the time.

7.9 Big Box Store Energy Consumption

The National Renewable Energy Laboratory (NREL) has analyzed the typical energy consumption (electricity and natural gas) from retail stores over different regions in the U.S. as summarized in Table 7.3 [4]. The average electricity consumption is fairly consistent across the US despite significant climate variations, primarily because indoor lighting accounts for approximately 51 % of all store electricity consumption. The average electricity consumption is 28 kwh per square foot per year, or approximately 2.792 million kWh/year for a 105,000 square foot building, which requires a PEM fuel cell system with approximately 330 kW of peak power capacity assuming a 96 % annual capacity factor, which is why we modeled a 330-kW fuel cell in this analysis.

7.10 Tri-Gen System Capital Cost Estimates

7.10.1 Stationary Fuel Cell System Cost

The low temperature PEM Fuel cell system cost estimates come from Brian James at Strategic Analysis Inc. (SAI) [5] James has been conducting detailed DFMA[10]

[10]DFMA = Design for Manufacturing and Assembly, a trademark of Boothroyd and Dewhurst.

cost estimates for the Department of Energy for many years, utilizing mass production cost estimation techniques that he learned under a mid-1990's DOE contract with the Ford Motor Company to assist them with analyzing the cost of hydrogen and fuel cell electric vehicle systems. James et al. analyzed the mass production costs of three types of stationary fuel cell systems, all running on natural gas:

- Low temperature PEM (proton Exchange Membrane) systems
- High temperature PEM systems, and
- Solid Oxide fuel cell (SOFC) systems

The SOFC system was the lowest cost, but we need to build a steam methane reformer (SMR) to convert natural gas to pure hydrogen to be used in fuel cell material handling equipment (MHE) such as forklift trucks and also in fuel cell electric vehicles (FCEVs), so we do not need the internal reforming system portion of the SOFC. We therefore chose a low temperature proton exchange membrane (PEM) fuel cell system, similar to the fuel cells used on MHE and FCEVs. The low-temperature PEM fuel cell has the added advantage of responding quickly to sudden increases in electricity demand.

James and his team considered primarily very high volume production runs of 100 to 50,000 units per year to represent vehicle manufacturing volumes, while we need cost estimates for much lower production volumes for early Tri-gen systems. James provided us with cost estimates for production volumes of only 10 units at power levels of 100 and 500 kWe as summarized in Table 7.4. [6] We then scaled between the 100 and 500 kWe numbers to arrive at an estimate of $2,071/kWe for the 330-kWe system required for the warehouses considered here. Note from Table 7.4 that these cost estimates all include a 10 % contingency above the estimated component costs, plus a 55 % markup that effectively converts the cost estimates into prices, plus a healthy installation cost addition of 28 % above the marked up price.

For the Tri-Gen system analyzed here, we also have to produce additional high purity hydrogen for use in the PEM fuel cells onboard FCEVs and fuel cell MHEs. This requires the addition of these components as shown in Fig. 7.1:

Table 7.4 SAI estimated cost of 100 and 500 kWe stationary PEM fuel cell systems with production quantities of 10 units

($/kWe)	100 kWe	330 kWe	500 kWe
FC Subsystem	$553/kW	$467/kW	$440/kW
Power electronics	$422/kW	$422/kW	$422/kW
System housing	$60/kW	$60/kW	$60/kW
Contingency (10 %)	$103.50	$94.88	$92.20
Sales margin (55 %)	$626.18	$574.04	$557.81
Installation (28 %)	$494.11	$452.97	$440.16
Total	$2259/kW	$2071/kW	$2012/kW

- Steam Methane Reformer (SMR) and pressure swing adsorption (PSA) gas clean-up system.
- Hydrogen compressor
- Hydrogen storage system
- Hydrogen refrigeration system (for 700-bar FCEVs only)
- Safety and controls
- Hydrogen dispenser(s)

The costs of these subsystems are analyzed in the following sections.

7.10.2 Steam Methane Reformer

We estimate the cost of a "Chemical plant" type SMR, such as the SMR technology developed by H2Gen Innovations, Inc. to be used at hydrogen fueling stations. This SMR system includes a built-in pressure swing adsorption (PSA) system that purifies the reformate from the SMR, reducing the CO content in the product hydrogen to levels suitable for use in a FCEV. We use the current cost estimates for SMR-PSA systems from an International Energy Agency report based on inputs from global experts in the areas of former manufacturers, gas companies and researchers[11] [7].

This estimate is part of Task 23 of the IEA Hydrogen Implementation Agreement.

7.10.3 Hydrogen Cascade Storage

Several organizations have shown that the least costly option to fill 700-bar FCEV hydrogen tanks is to use a cascade storage system. Rather than compressing all hydrogen to pressures above 700 bar to fill 700-bar tanks, a series of hydrogen storage tanks with increasing pressure are provided at the station. When a FCEV arrives at the station, the car tanks are first attached to the lowest pressure storage tank. When the car tank pressure reaches the pressure of the lowest pressure in the cascade, then controls automatically connect the car tank to the next higher pressure cascade, until the highest pressure tanks top off the FCEV tank to 700 bar. This cascade filling approach allows some hydrogen to fill car tanks without having to be pressurized to the highest pressure above 700-bar, saving on compressor cost and compression electricity costs. Although fuel cell forklift trucks generally use lower pressure 350-bar hydrogen, we include the cost of the 700-bar cascade filling system for FCEVs to be conservative.

[11]The logarithmic curve fit shown on Fig. 7.8 goes negative for less than 60 kg/day capacity, so a separate polynomial curve fit had to be developed for small SMR systems.

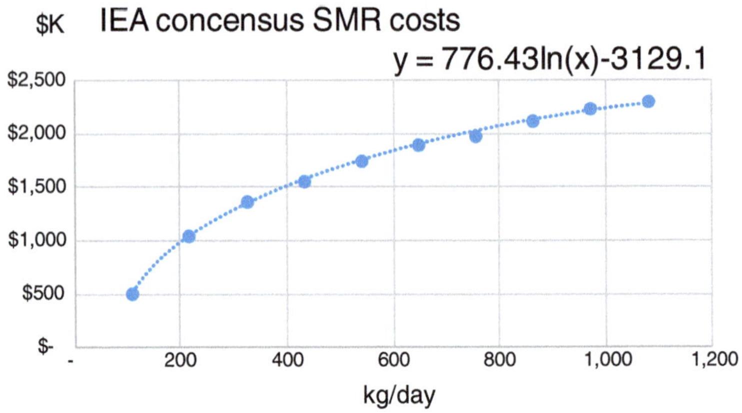

Fig. 7.8 Estimated installed price for an SMR-PSA system as a function of capacity in kg/day in low production volumes

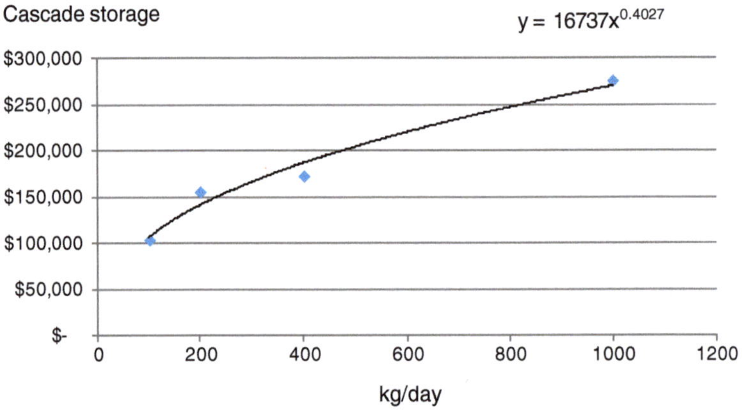

Fig. 7.9 Estimated cost of a high pressure (700-bar) hydrogen cascade storage system

K. Reddi et al. at the Argonne National Laboratory (ANL) have shown that the optimum cascade system is a four-level cascade, with each level comprising a single hydrogen tank [8]. Their cost estimates (Fig. 7.9) include site preparation, engineering and design, permitting and installation.[12]

[12]While the hydrogen storage tanks on fuel cell forklift trucks are generally lower pressure (350-bar), we include the higher pressure (700-bar) cascade storage system to accommodate FCEVs.

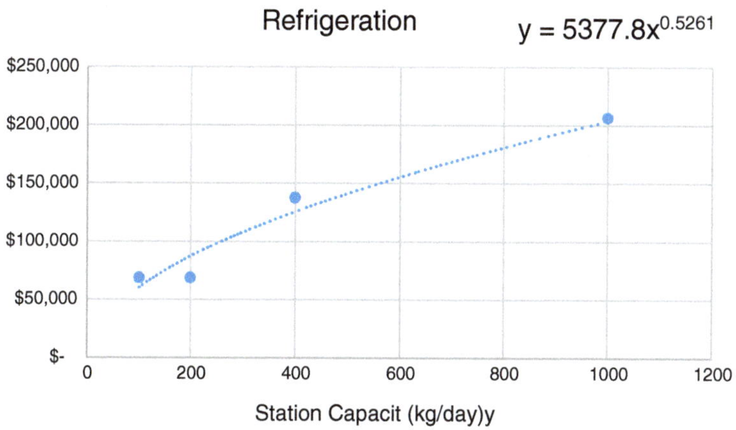

Fig. 7.10 Estimated cost of hydrogen refrigeration system as a function of station capacity

7.10.4 Hydrogen Refrigeration

Hydrogen gas temperature rises during compression. If the hydrogen in the tanks is heated, then it may not be possible to reach the design pressure of 700-bar, which would decrease the amount of hydrogen that can be stored and hence would reduce the range of the FCEV. To provide a full tank of hydrogen, the hydrogen is precooled to a temperature near −40 °C. Amgad Elgowainy and his team at ANL [9] have estimated the cost of refrigeration as a function of the station capacity as shown in Fig. 7.10. Refrigeration costs vary from approximately $53,000 for a 100 kg/day station to $211,000 for a 1,000 kg/day station.

7.10.5 700-Bar Compression

To estimate the installed price of a 700-bar compressor, we used the estimates from Elgowainy at the Argonne National Laboratory[13] [9]. The final installed price estimate for a 700-bar compressor is shown in Fig. 7.11 as a function of station capacity in kg/day. To be conservative, we add the cost for *two* compressors per station, since compressors are notoriously unreliable.

---

[13]We compared the ANL compressor cost estimates with three vendors and two other national laboratories. In all cases the ANL price estimates were higher than the others.

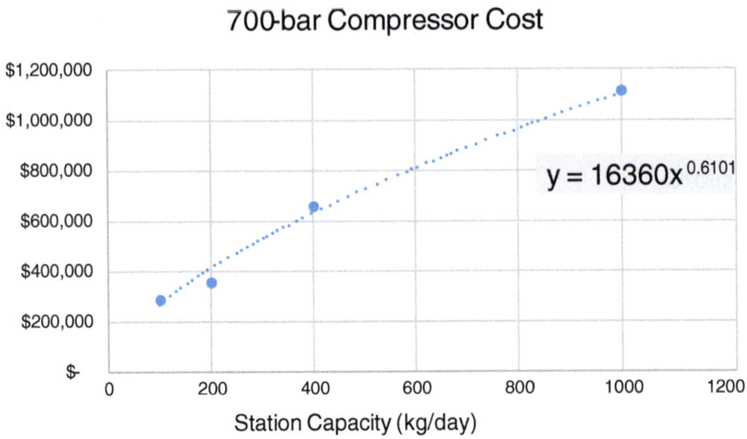

Fig. 7.11 estimated installed price of a 700-bar compressor as a function of station capacity in kg/day

7.10.6 CSD Controls and Safety

The cost of compression, storage and dispensing (CSD) controls along with safety devices was estimated by the Argonne National Laboratory at $144,000 independent of station capacity [9].

7.10.7 Dispenser

Dispenser costs in the literature have been underestimated according to a DOE technical team peer review of hydrogen station cost estimates [10]. While the main DOE H2A cost model assigned a cost of $53,667 per dispenser, the independent review team estimated an average cost of approximately $98,444 per dispenser, with one two-hose dispenser required for every 250 kg/day of station capacity.

7.10.8 Other Capital Costs

In their analysis of station costs, the Argonne National Laboratory also listed another category of undefined "other capital" costs [9]. These added costs increase with station capacity as shown in Fig. 7.12.

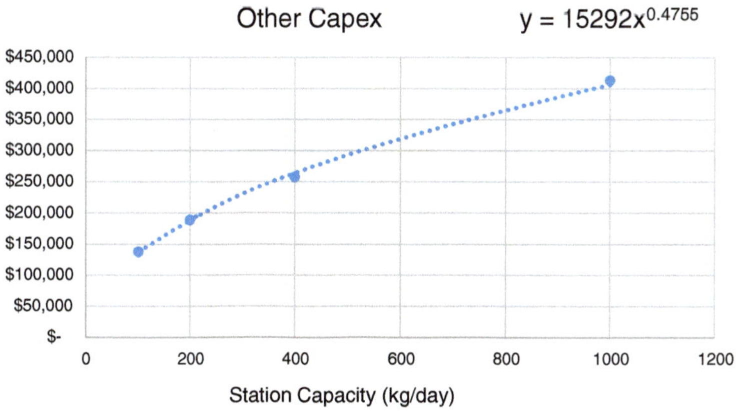

Fig. 7.12 hydrogen station "Other costs" from the Argonne National Laboratory

7.11 Summary Station Costs

The total estimated costs for the Tri-Gen system supporting 10 FCEVs are shown in Table 7.5 as a function of the number of fuel cell fork lift trucks supported. The SMR system capacity must be increased for more than 100 MHE trucks.

7.11.1 Cash Flow Chart

Table 7.6 on the next page shows the cash flows for the first 15 years for the California case with 10 FCEVs using the station along with 100 fuel cell fork lift

Table 7.5 Tri-Gen System installed prices as a function of the number of fuel cell fork lift trucks supported (including enough hydrogen for 10 FCEVs)

# of Fork lift trucks =>	50–100	150	200
PEM FC system	$698,058	$698,058	$698,058
SMR + PSA	$241,707	$420,803	$594,634
Storage	$303,528	$303,528	$303,528
Refrigeration	$74,378	$74,378	$74,378
Compressor	$476,245	$476,245	$476,245
Electrical	$75,238	$75,238	$75,238
Dispenser	$68,440	$68,440	$68,440
Controls and safety	$144,000	$144,000	$144,000
"Other" capital cost	$142,614	$142,614	$142,614
Total capex	$2,224,209	$2,403,304	$2,577,135

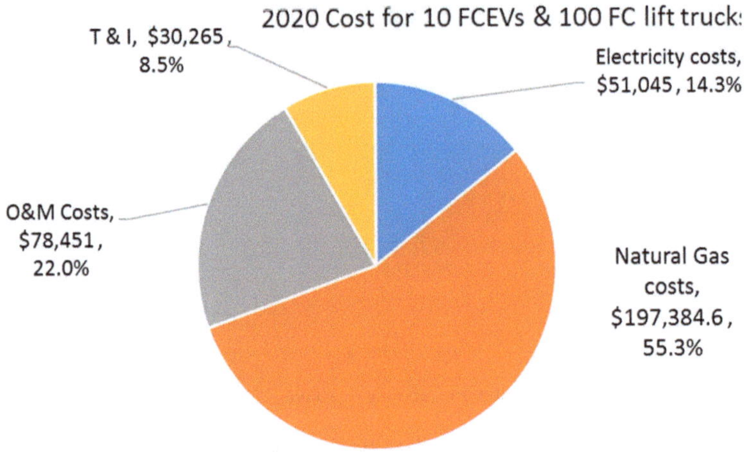

Fig. 7.13 Estimated costs for a station with 100 material handling equipment (MHE) and 10 FCEVs in California in 2020

trucks, which provides a 20-year IRR of 18.5 %, and a 30-year IRR of 19.1 %. We assume an eight-month construction time, so costs and revenues are reduced by 66 % in the first year of the project.

The annual costs in 2020 for the California case are shown in Fig. 7.13. Natural gas dominates that cost at 55.3 %; purchased electricity accounts for 14 % of costs. The station must buy electricity for 8 % of the time, since we assume a 92 % capacity factor. Taxes and insurance (T&I) account for 9 % of annual costs, and operation and maintenance accounts for 22 % of all costs.

Similarly, Fig. 7.14 shows the station revenues for 2020 in California, assuming that 10 FCEVs and 100 lift trucks are using the station. Avoided electricity costs dominate the "revenue" at 43 % with hydrogen revenue from 10 FCEVs supplying only 2 % of the revenues and heat adding less than 1 % from avoided heating costs (not shown). The "avoided LH2 infrastructure costs" are due to avoiding the monthly charge of $17,000 from the industrial gas companies for leasing the liquid hydrogen on-site system.

Table 7.6 Cash flow statement for first 15 years for a tri-gen system in California with 10 FCEVs and 100 fuel cell fork lift trucks

		2015	2016	2017	2018	2019	2020	2021	2022
Gasoline price	$/gallon	2.69	3.06	3.20	3.22	3.24	3.28	3.33	3.36
Comm NG $/MBTU	commercial	14.86	14.88	15.00	14.97	15.61	16.26	16.72	16.97
IN NG $/MBTU	Industrial	5.26	5.37	5.65	5.96	6.32	6.67	6.91	7.04
NG for electric power ($/MBTU)	Elect Power	4.35	4.35	4.38	4.41	4.77	5.10	5.37	5.53
NG price	$/MBTU(HHV)	6.89	7.03	7.40	7.81	8.27	8.73	9.06	9.22
H2 Eq. price	$/kg	6.56	7.469	7.81	7.84	7.89	8.00	8.11	8.20
H2 sales to FCEVs	kg/year	555	1664	1664	1664	1664	1664	1664	1664
Electricity-comm	cents/kWh	15.98	16.56	16.67	16.37	16.25	16.27	16.31	16.32
Commercial elect	cents/kWh	15.98	16.56	16.67	16.37	16.25	16.27	16.31	16.32
Revenue	multiplier	0.333	1	1	1	1	1	1	1
Electricity avoided	MWh/yr	887	2660	2660	2660	2660	2660	2660	2660
Purchased electricity	MWh/yr	77	231	231	231	231	231	231	231
SMR electricity	MWh/yr	11	33	33	33	33	33	33	33
400-bar compr. electri	MWh/yr	15	45	45	45	45	45	45	45
900-bar compr. electricty	MWh/yr	2	5	5	5	5	5	5	5
Elect revenue (avoided costs)	$K	$47	$440	$443	$435	$432	$433	$434	$434
Avoided LH2 costs	$K	$56	$169	$169	$169	$169	$169	$169	$169
Avoided H2 infrastructure cost	$K	$68	$204	$204	$204	$204	$204	$204	$204
H2 revenue-FCEVs	$K	$1	$12	$13	$ 13	$ 13	$13	$13	$14
Heat use	MBTU/yr	$123	$368	$368	$368	$368	$368	$368	$368
Heat revenue		$0.197	$1.809	$1.907	$2.010	$2.130	$2.249	$2.332	$2.375
Total revenue		$173	$827	$ 831	$823	$820	$821	$822	$823
Capex	$K	$2,208.28	$						
Construction finance charges	$K	$265.0	$						

(continued)

Table 7.6 (continued)

		2015	2016	2017	2018	2019	2020	2021	2022
O&M	$K	$30	$90	$90	$90	$ 90	$90	$90	$90
Purchased electricity	$K	$17	52	$52	$51	$ 51	$51	$51	$51
Taxes and Ins.	1.5 %	$11.04	$33.12	$33.12	$33.12	$33.12	$33.12	$33.12	$33.12
NG req'd for SOFC	MBTU/yr	2057.7	18,519.1	18,519.1	18,519.1	18,519.1	18,519.1	18,519.1	18,519.1
NG req'd for MHE	MBTU/yr	1261.3	3783.8	3783.8	3783.8	3783.8	3783.8	3783.8	3783.8
NG req'd for FCEVs	MBTU/yr	99.6	298.7	298.7	298.7	298.7	298.7	298.7	298.7
NG Cost	$K/yr	$23.6	$158.8	$167.3	$176.4	$187.0	$197.4	$204.7	$208.5
Total costs		$2,555	$334	$343	$351	$361	$371	$379	$383
Cash flow		−2382	493	488	472	459	449	443	440
Cal	MHE Qty=	100							
15-yr IRR	17.2 %								
20-yr IRR	18.5 %								
30-yr IRR	19.1 %								
		2023	2024	2025	2026	2027	2028	2029	2030
Gasoline price	$/gallon	3.41	3.46	3.52	3.57	3.62	3.69	3.76	3.82
Comm NG $/MBTU	commercial	17.24	17.38	17.75	18.08	18.10	17.93	17.90	17.95
IN NG $/MBTU	Industrial	7.19	7.25	7.47	7.66	7.57	7.36	7.26	7.20
NG for electric power ($/MBTU)	Elect Power	5.69	5.78	6.04	6.23	6.22	6.09	6.06	6.08
NG price	$/MBTU(HHV)	9.41	9.49	9.79	10.03	9.91	9.64	9.50	9.43
H2 Eq. price	$/kg	8.32	8.44	8.57	8.70	8.83	9.00	9.16	9.31
H2 sales to FCEVs	kg/year	1664	1664	1664	1664	1664	1664	1664	1664
Elecricity -comm	cents/kWh	16.22	16.14	16.19	16.27	16.28	16.16	16.02	15.98
Commercial elect	cents/kWh	16.22	16.14	16.19	16.27	16.28	16.16	16.02	15.98

(continued)

Table 7.6 (continued)

		2023	2024	2025	2026	2027	2028	2029	2030
Revenue	multiplier	1	1	1	1	1	1	1	1
Electricity avoided	MWh/yr	2660	2660	2660	2660	2660	2660	2660	2660
Purchased electricity	MWh/yr	231	231	231	231	231	231	231	231
SMR electricity	MWh/yr	33	33	33	33	33	33	33	33
400-bar compr. electri	MWh/yr	45	45	45	45	45	45	45	45
900-bar compr. electricty	MWh/yr	5	5	5	5	5	5	5	5
Elect revenue (avoided costs)	$K	$431	$429	$431	$433	$433	$430	$426	$425
Avoided LH2 costs	$K	$169	$169	$169	$169	$169	$169	$169	$169
Avoided H2 infrastructure cost	$K	$204	$204	$204	$204	$204	$204	$204	$204
H2 revenue-FCEVs	$K	$14	$14	$14	$ 14	$15	$15	$15	$15
Heat use	MBTU/yr	$368	$368	$368	$368	$368	$368	$368	$368
Heat revenue		$2.424	$2.445	$2.520	$2.584	$2.552	$2.482	$2.447	$2.429
Total revenue		$820	$818	$820	$822	$823	$820	$816	$816
Capex	$K								
Construction finance charges	$K								
O&M	$K	$90	$90	$90	$ 90	$90	$90	$90	$90
Purchased electricity	$K	$51	$51	$51	$ 51	$51	$51	$50	$50
Taxes and Ins.	1.5 %	$33.12	$33.12	$33.12	$33.12	$33.12	$ 33.12	$33.12	$33.12
NG req'd for SOFC	MBTU/yr	18,519.1	18,519.1	18,519.1	18,519.1	18,519.1	18,519.1	18,519.1	18,519.1
NG req'd for MHE	MBTU/yr	3783.8	3783.8	3783.8	3783.8	3783.8	3783.8	3783.8	3783.8
NG req'd for FCEVs	MBTU/yr	298.7	298.7	298.7	298.7	298.7	298.7	298.7	298.7

(continued)

Table 7.6 (continued)

		2023	2024	2025	2026	2027	2028	2029	2030
NG Cost	$K/yr	$212.7	$214.6	$221.2	$226.8	$223.9	$ 217.9	$214.8	$213.2
Total costs		$387	$388	$395	$401	$398	$392	$388	$386
Cash flow		433	430	425	422	425	428	428	429
Cal	MHE Qty=								
15-yr IRR	17.2 %								
20-yr IRR	18.5 %								
30-yr IRR	19.1 %								

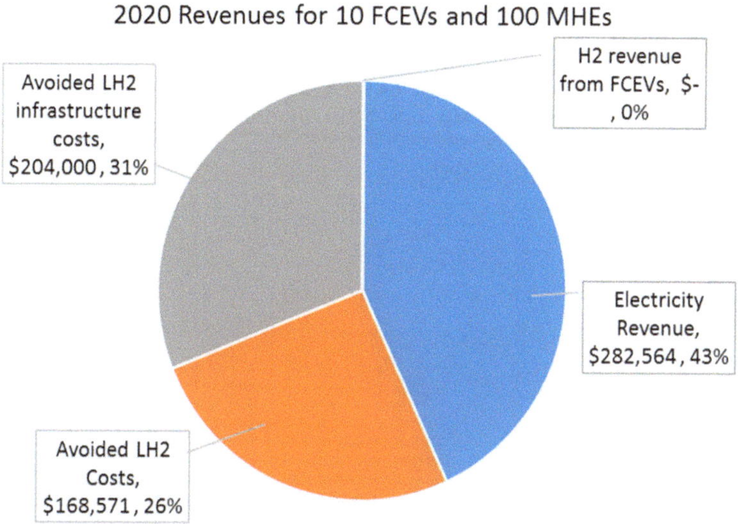

Fig. 7.14 Pie chart of 2020 Revenues in California with 10 FCEVs and 100 fuel cell lift trucks using the station

References

1. Department of Energy Electricity Table Browser. Available at:http://www.eia.gov/electricity/data/browser/#/topic/7?agg=0,1&geo=vvvvvvvvvvvvo&endsec=vg&linechart=ELEC.PRICE.US-ALL.A&columnchart=ELEC.PRICE.US-ALL.A&map=ELEC.PRICE.US-ALL.A&freq=A&ctype=linechart&ltype=pin&rtype=s&pin=&rse=0&maptype=0
2. Natural Gas prices by state, U.S. Energy Information Administration. Available at: http://www.eia.gov/dnav/ng/ng_pri_sum_a_EPG0_PEU_DMcf_a.htm
3. Edelstein S (2014) FedEx follows French Lead, Tests hydrogen fuel-cell range extenders. Green Car Reports, 27 Jan 2014. Available at: http://www.greencarreports.com/news/1089920_fedex-follows-french-lead-tests-hydrogen-fuel-cell-range-extenders
4. Bonnema E et al (2013) Development of the advanced energy design guide for medium to big box retail buildings—50 % energy savings. The National Renewable Energy Laboratory, NREL/TP-5500-52589, June 2013
5. James BD et al (2012) Manufacturing cost analysis of stationary fuel cell systems, Rev 4. Strategic Analysis, Inc., Arlington, Virginia, September 7, 2012. Available at: https://www.sainc.com/service/SA%202012%20Manufacturing%20Cost%20Analysis%20of%20Stationary%20Fuel%20Cell%20Systems.pdf
6. James BD (2016) private communication, 5/9/2016
7. Schjolberg I et al (2012) Small-scale reformers for on-site hydrogen supply. Energy Procedia 29:559–566
8. Reddi K et al (2014) the Argonne National Laboratory. Hydrogen refueling station compression and storage optimization with tube-trailer deliveries. Int J Hydrogen Energy (39):pgs 19169–191181

9. Elgowainy A et al (2012) Hydrogen delivery and infrastructure analysis, DOE Hydrogen program review, May 17, 2012. Available at: https://www.hydrogen.energy.gov/pdfs/review12/pd014_mintz_2012_o.pdf
10. Parks G et al (2014) DOE Independent Review Team. Hydrogen station compression, storage and dispensing technical status and costs. NREL/BK-6A10-58562, May 2014

Chapter 8
Economics of Hydrogen and Electricity Co-generation at IGCC Plants

Abstract Co-producing hydrogen and electricity at coal-powered integrated gasification combined cycle (IGCC) plants with carbon capture and storage (CCS) provides high return on the investment by selling two products while enabling the US to meet the greenhouse gas (GHG) reduction goal of 80 % below 1990 levels. Thirty-year Internal rates of return above 10–15 % are feasible with these co-generation plants, assuming that they sell electricity to the grid and hydrogen to local liquid hydrogen dispensing stations for fuel cell electric vehicles or fuel cell forklift trucks.

8.1 Introduction

We have shown in Chap. 3 that the U.S. can achieve our societal goal of cutting GHGs to 80 % below 1990 levels by 2050 based primarily on making most electricity from coal-fired integrated gasification combined cycle (IGCC) plants with carbon capture and storage (CCS). This chapter addresses the likely return on investment on these plants, assuming that they sell both electricity and hydrogen. We assume that fuel cell electric vehicles (FCEVs) enter the light duty vehicle (LDV) fleet in Southern California and provide a significant market for hydrogen from an IGCC+CCS plant installed east of Los Angeles, where the hydrogen is liquefied and shipped by tanker truck to the LA region. Rates of return are also high in Hawaii due primarily to much higher electricity prices, but California is more aggressive in terms of installing hydrogen stations and promoting the early introduction of FCEVs, providing an earlier profitable venue for IGCC+CCS plants.

C.E. Thomas, *Stopping Climate Change: the Case for Hydrogen and Coal*,
Lecture Notes in Energy 35, DOI 10.1007/978-3-319-31655-0_8

Table 8.1 Estimated performance and costs for coal-powered IGCC+CCS plants co-producing hydrogen and electricity; Costs are in millions of dollars or euros

	IEA-Mix	IAE-Lo H2	IEA-Hi H2
Net electricity output (MW)	317.0	363.1	236.6
Coal input (MW)	1963	1963	1963
Hydrogen output (MW)	599	484	734.1
Hydrogen output (kg/day)	433,277	349,969	530,994
Cost (2008 US-€M)	1672	1689	1689
Cost (2008 US$M)	$2508	$2534	$2534
Cost (2015 US$M)	$2470	$2495	$2495
Fixed O&M cost (2015US$M)	$68.37	$69.09	$69.09

8.2 Capital Costs of Electricity and Hydrogen Co-generation at IGCC+CCS Plants

A 2008 International Energy Agency (IEA) report based on detailed analysis and calculations by the global engineering firm Foster-Wheeler, estimates the capital costs for IGCC+CCS plants with several different ratios of hydrogen to electricity production as shown in Table 8.1 [1]. We have used the exchange rate of 1 € = $1.50 at the time (2008) of these estimates, and have converted the resulting 2008$ to 2015$ using the North American Power Capital Cost Index (PCCI) [2].[1]

8.3 Capital Cost of Liquid Hydrogen Production

The least costly option for delivering hydrogen in the early days of market penetration before hydrogen consumption is large enough to justify building hydrogen pipelines is to liquefy the hydrogen at central plants and deliver it by cryogenic tanker trucks to the end user.[2] Liquid hydrogen tanker trucks routinely supply liquid hydrogen several thousand miles from several liquid hydrogen plants on the East Coast of Canada in Ontario and Quebec to hydrogen and fuel cell development companies on the West coast around Vancouver, British Columbia. Liquefaction plants are expensive and also consume significant electricity to liquefy the hydrogen. But Yang and Ogden have shown that liquid hydrogen is the lowest cost alternative to bring hydrogen to fueling stations before hydrogen pipelines are built [3], despite the high capital and electricity costs of liquefying hydrogen.

[1]The PCCI index for power plants fell slightly to 186 in 2015, compared to 189 in 2008, after hitting 190 in 2014, so the PCCI adjustment slightly *reduces* the estimated 2015 costs compared to 2008 costs.

[2]This was the delivery method used to supply liquid hydrogen for the space program beginning in the 1950s and is used today to deliver liquid hydrogen to many warehouses with hydrogen-powered fuel cell material handling equipment such as fork-lift trucks.

Table 8.2 Estimated capital costs for hydrogen liquefaction plants in 2015 million dollars

	30 tonne/day	300 tonne/day
Praxair	$102.7	$788.0
Air Products	$78.5	$272.0
NREL	$56.2	$255.2
Nexant Inc.	$77.1	$342.63
Average	$78.6	$414.5

Table 8.3 Hydrogen capacity and capital cost in million U.S. 2015 dollars for the three IEA cases

H2	LH2 capex
kg/day	(2015 U.S.$ millions)
431,530	525
348,682	457
528,859	598

Table 8.4 Total capital costs for IGCC+CCS plants, LH2 liquefaction plants and a fleet of LH2 tanker trucks

	IGCC+CCS plant	LH2 plant	LH2 tankers	Total
IEA-Mix	$2,470,480,913	$524,548,721	$495,600,000	$3,490,629,634
IEA-Lo-H2	$2,511,264,050	$456,870,426	$400,800,000	$3,368,934,476
IEA-Hi H2	$2,494,927,481	$598,442,787	$607,800,000	$3,701,170,267

To estimate the cost of a hydrogen liquefaction plant (including on-site storage for 5 days of LH2 production), we use the average of four sources: industrial gas companies (Praxair [4] and Air Products [5]) from the 1995–1996 time period, Ramsden et al. from the National Renewable Energy Laboratory (NREL) [6] and a presentation by Nexant, Inc. [7] as summarized in Table 8.2. We escalated the 1995 costs to 2015$ by a factor of 2.41 from the PCCI data [2] to account for the much higher capital costs for power plants in 2015 compared to 1995. In each case we scaled the liquefaction plants to match the hydrogen production capacity specified in the three cases in the IEA report as summarized in Table 8.3.

The total capital costs for the IGCC+CCS plants plus the liquefaction plant and enough liquid hydrogen tanker trucks to deliver the hydrogen is summarized in Table 8.4. The total costs per plant exceed $3.3 billion.

One possible option to maximize hydrogen sales might be to build an IGCC +CCS plant in a region with multiple warehouses that are already using hydrogen-powered fuel cell material handling equipment.[3] Another option would

[3]Although, as described above, the costs to transport liquid hydrogen hundreds or even thousands of mile is manageable.

be to place the IGCC+CCS plant close to industrial facilities that consume large quantities of hydrogen.

In this chapter we assume that the State of California continues their plan to introduce thousands of FCEVs in the next few years. We assume that an IGCC +CCS plant is built east of Los Angeles to supply liquid hydrogen to multiple hydrogen fueling stations in the LA basin.

8.4 Potential FCEV Markets in California and Hawaii

To estimate the return on investment for IGCC+CCS plants, we assume a gradual introduction of FCEVs in Southern California as shown in Fig. 8.1. There were 6.197 million vehicles registered in Los Angeles County in 2014 [8]. We represent the number of FCEVs in the LA region by a logistics curve:

$$N = A/[(1 + \exp(-k \times (T - T_o))]$$

where A = the total number of vehicles (6.197 million),

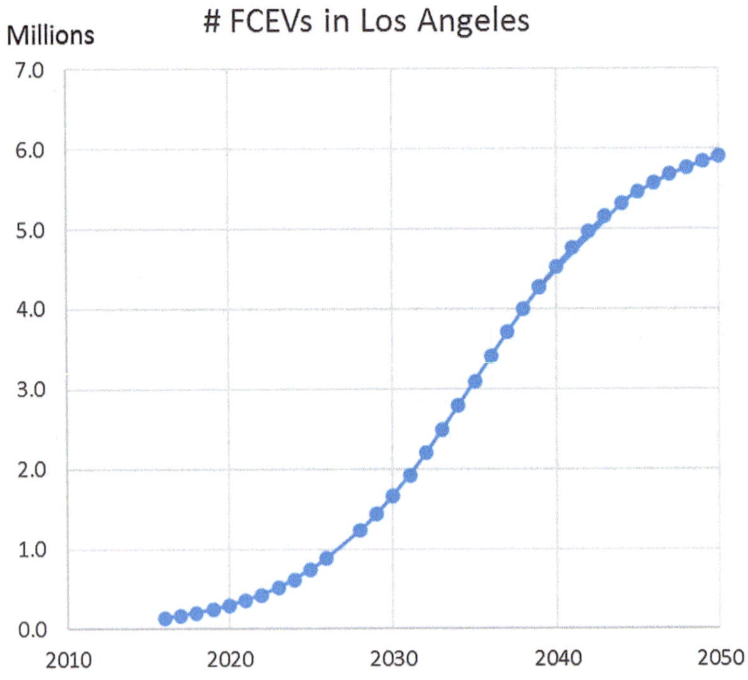

Fig. 8.1 Assumed number of fuel cell electric vehicles on the road in the Los Angeles region over time

K = 0.2,
T= current year, and
T_o = 2035 = year with 50 % FCEVs.

Thus we assume that half of all light duty vehicles on the road are FCEVs by 2035 as shown in Fig. 8.1.

8.5 Hydrogen Consumed by FCEVs

We need to calculate the hydrogen consumed each year by a fuel cell electric vehicle (FCEV) to determine how much hydrogen to produce at the IGCC+CCS plant, and also to estimate how many liquid hydrogen (LH2) fueling stations must be built to accommodate the FCEVs. The hydrogen consumed by a FCEV each year will depend on the fuel economy of the FCEV and the vehicle miles traveled (VMT) each year. The average fuel economy of the early prototype FCEVs introduced in the early 2000s was 61.4 miles per kg of hydrogen and shown in Table 6.6 in Chap. 6. Assuming the 2013 average U.S. VMT of 10,220 miles per year [9], then a FCEV would consume on the average:

$$10{,}220/61.4 = 166\,\text{kg/year or approximately}\,0.46\,\text{kg/day}$$

8.6 Setting the Price of Hydrogen

We set the price of hydrogen such that the driver of a FCEV pays the same cost for hydrogen *per mile* as the driver of a conventional vehicle pays for gasoline per mile traveled. The average 2013 U.S. vehicle had a fuel economy of 25.2 miles per gallon of gasoline[4] [9]. If gasoline costs \$3.30/gallon, then the driver pays:

$$\$3.30/(25.2 \times 100) = 13.1\ \text{¢/mile.}$$

The price of hydrogen for a FCEV with 61.4 miles/kg fuel economy would then be:

$$13.1 \times 61.4/100 = \$8.05/\text{kg}.$$

[4]The fuel economy of gasoline vehicles will increase in the future as the auto companies are required to meet new fuel economy standards; thus most new gasoline vehicles will have to be hybrid electric vehicles (HEVs) to meet these new standards. But the fuel economy of FCEVs will also undoubtedly increase over the 61.4 miles per kg estimate from Chap. 6, since many of the factors that will increase conventional vehicle fuel economy will also improve the fuel economy of FCEVs (reduced aerodynamic drag, reduced body weight, reduced cross-sectional area, etc.) In addition, the efficiency of the fuel cell system itself will most likely increase over time.

Table 8.5 Estimated installed cost of one hydrogen fueling station receiving liquid hydrogen and dispensing gaseous hydrogen at 700-bar for FCEVs

LH2 station capacity (kg/day)	LH2 storage	LH2 pump	Vaporizer	GH2 cascade	Dispensers (1–3)	8 % site prep and permits	8 % markup	10 % eng. and design	5 % contingency	30 % installation	Total installed cost
150	$41,049	$16,688	$14,336	$47,453	$242,274	$28,944	$31,259	$42,200	$23,210	$146,224	$633,637
300	$66,684	$32,747	$26,164	$94,907	$242,274	$37,022	$39,984	$53,978	$29,688	$187,034	$810,479
400	$81,560	$43,362	$34,049	$126,542	$242,274	$42,223	$45,601	$61,561	$33,859	$213,309	$924,339
1000	$154,894	$106,408	$81,359	$316,355	$484,547	$91,485	$98,804	$133,385	$73,362	$462,180	$2,002,778
2000	$251,626	$210,444	$160,209	$632,711	$726,821	$158,545	$171,228	$231,158	$127,137	$800,964	$3,470,842

So the final price of hydrogen is proportional to the cost of gasoline, given by:

$$\text{Hydrogen Price} = \$8.05/\text{kg} \times [\text{Gasoline price in } \$/\text{gallon}]/3.30.$$

8.7 Liquid Hydrogen Fueling Station Cost

We assume that the owner of the IGCC+CCS plant liquefies the hydrogen and delivers liquid hydrogen to hydrogen fueling stations. Other entities such as the industrial gas companies, possibly the automobile companies or other private companies might also build or finance these LH2 stations. But we assume that the owner of the IGCC+CCS plant also builds, installs, and operates the LH2 stations necessary to supply hydrogen to the FCEVs.

We estimate the cost of these LH2 fueling stations based on a detailed analysis by Yang and Ogden who summarized cost estimates from several other sources [10]. The total estimated station costs are summarized in Table 8.5 as a function of station capacity. We have escalated the 2007 cost estimates by a factor of 1.05 to account for inflation in 2015 dollars according to the Power Capital Cost Index (PCCI) [2]. The estimated LH2 station cost is $633,637 in 2015$ for a small LH2 station dispensing only 150 kg/day up to $3.47 million for a larger station with a capacity of 2000 kg/year. This larger station can support approximately 4100 FCEVs assuming an average consumption rate of 0.46 kg/day, a station capacity factor of 92 % and annual VMT of 10,220 miles per vehicle.

In the model, LH2 fueling stations are added each year to support the number of new FCEVs introduced in the *following* year, so the capital expenses for these stations are incurred the year before any revenue is generated from selling hydrogen to FCEVs. Each IGCC+CCS plant can support between 727,000 (IEA Lo-H_2 Case) and 1.11 million (IEA Hi-H_2 Case) FCEVs. Once a plant reaches capacity, another IGCC+CCS plant must be built the previous year along with a set of LH2 stations to dispense the hydrogen from that plant or LH2 must be trucked in from outside plants to meet the hydrogen demand of FCEVs.

8.8 Impact of Carbon Tax on IGCC+CCS Electricity Margin

While the likelihood of carbon tax legislation is remote with the current U.S. congress, we assume here that carbon taxes are eventually passed as the impact of climate change forces voters to insist that Congress act.

Table 8.6 Estimated net revenue for three power plant types for the first year of the Citizen's Climate Lobby proposed carbon fees, assuming a market clearing price of 10 ₵/kWh

	Clearing price (₵/kWh)	GHGs (g/kWh)	Carbon tax ($/tonne)	(₵/kWh) (c/kWh)	Net generator "revenue"
Coal-ST	10.00	1115.3	$16.54/tonne	1.84	8.16
NGCC	10.00	507.2	$16.54/tonne	0.84	9.16
IGCC+CCS	10.00	87.4	$16.54/tonne	0.14	9.86

Although any carbon tax will certainly be passed on to the consumer, any tax will still have a small positive impact on the owner of an IGCC+CCS plant. Since most electricity in the U.S. is now sold in competitive wholesale markets, the price of electricity is set by the highest cost bidder necessary to meet the next day's load. For the next few decades the market-based clearing price will be set by some combination of coal steam turbine plants and natural gas combined cycle (NGCC) plants. Both of these plants generate more greenhouse gases (GHGs) than an IGCC +CCS plant. The bid price (assumed to be 10 ₵/kWh here) includes the carbon costs as shown in Table 8.6. We assume that the clearing price is set by an average of the prices from Coal-ST and NGCC plants.[5]

Since the owner of the IGCC+CCS plants receives the same clearing price as the more polluting plants, the IGCC+CCS owner net margin (after subtracting its carbon cost) is higher than the other two plants that the owner can use to pay for operating costs. The differences are very small, on the order of a fraction of a ₵/kWh to at most 1.7 ₵/kWh the first year, but his small incremental increase in revenue can make a difference in rates of return. In addition, the carbon taxes increase over time as discussed in Appendix A, which increases the rates of return.[6] We also analyze the case without any carbon tax in the last section of this chapter, showing that 10–13 % 30-year returns are possible without any carbon tax.

8.9 Return on Investment for IGCC+CCS Plants

The calculated 20-year and 30-year IRRs for these plants are summarized in Table 8.7 for the three IEA cases in the Los Angeles region. Recall that these returns are on investments of more than $3.3 billion as shown in Table 8.4, which

[5]See Appendix A for a description of the various carbon tax plans.

[6]The $16.54/tonne tax shown in Table 8.5 is for the first year of the Citizen's Climate Lobby proposal.

Table 8.7 Returns on investment for the three IEA IGCC+CCS plants for the WECC-California fuel prices

	20-year IRR (%)	30-year IRR (%)
IEA-Mix	10.30	12.32
IEA-Lo H2	10.76	12.68
IEA-Hi H2	8.72	11.08

Table 8.8 Impact of reduced hydrogen prices on rates of return in LA

Ratio H2 price to gasoline price (%)	15-year IRR (%)	20-year IRR (%)	30-year IRR (%)
70	2.9	6.5	9.2
80	4.8	8.07	10.47
90	6.5	9.48	11.61
100	8.0	10.76	12.68

does not include the costs over time of installing the LH2 stations at a cost of \$3.47 million each.[7]

Hydrogen Price Discount. The IRRs shown in Table 8.7 assume that the price of hydrogen per mile driven in a FCEV is set equal to the price of gasoline per mile driven in a conventional vehicle. The LH2 plant owner could reduce the hydrogen price and still make a reasonable rate of return as shown in Table 8.8; thus the owner could sell hydrogen at a 20 % discount to gasoline per mile driven and still make a 30-year return above 10 %.

Hawaii. Hawaii might seem to be an improbable place to build a coal-powered IGCC+CCS plant, since Hawaii has no known coal reserves. However, Hawaii has been operating a single 180-MW coal-powered generation plant on Oahu since 1993 [11]. The coal is imported from Indonesia, unloaded at a dock, and transported to the power plant by a 1.6-mile conveyor belt. Coal is considered a welcome diversification of electricity fuel supply in Hawaii, since 78 % of electricity on Hawaii is generated from petroleum. The coal reserves in Indonesia are substantial, with an annual production to reserve ratio of 61 [12].

Hawaii (Oahu) has far fewer vehicles (≈816,000) than Los Angeles (6.197 million). We assume that FCEV sales on Oahu, like LA, reach 50 % by 2035 as shown in Fig. 8.2.

Despite having to import all coal, Hawaiian coal costs only twice the cost in California and the rest of the US as summarized in Table 8.9 for 2016. Counterbalancing the high cost of coal, the cost of electricity on Hawaii is the highest in the nation, and therefore will generate more revenue from selling electricity from the IGCC+CCS plant. On the downside, commercial electricity prices are also high in Hawaii, which will slightly increase operating costs for the LH2

[7]Although the costs of the liquid hydrogen stations are not included in Table 8.4, they are included in the cash flow statement used to calculate IRRs.

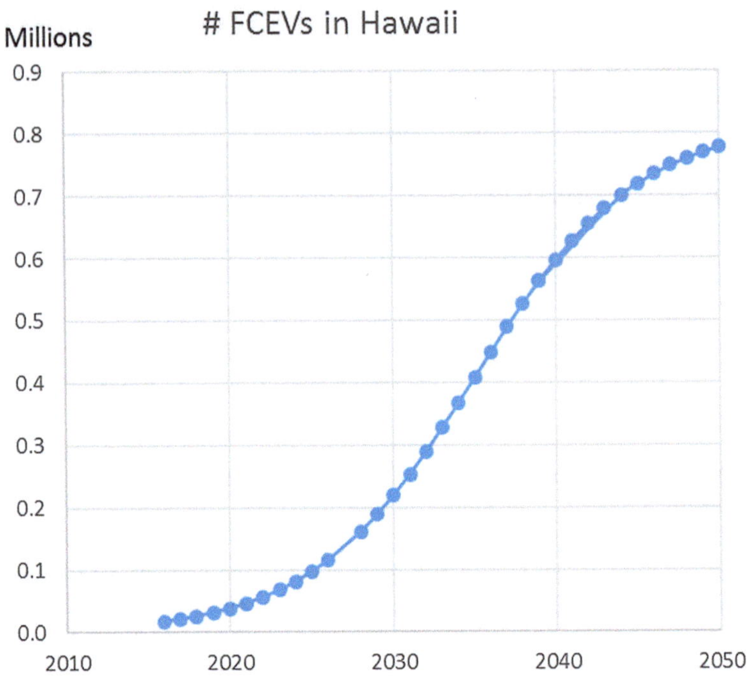

Fig. 8.2 FCEV on the road assumed on Oahu

Table 8.9 2016 fuel cost comparisons for Hawaii, California and the US average

	Hawaii	California	US average
Coal ($/MBTU)	4.06	1.97	2.37
Gasoline ($/gallon)	2.58	3.03	2.49
Com. electricity (₵/kWh)	34.60	14.76	10.36
All-sector electricity	33.62	14.87	10.30

fueling stations relative to the mainland. However, the slightly increased LH2 station operating costs are more than offset by the much higher revenues from selling high price electricity into the general marketplace in Hawaii.

Table 8.10 Rates of return for the three IEA cases on Hawaii

	20-year IRR (%)	30-year IRR (%)
IEA-Mix	10.28	12.13
IEA-Lo H2	13.40	14.79
IEA-Hi H2	3.92	7.05

Table 8.11 Rates of return on Hawaii with hydrogen prices below gasoline prices per mile driven for IEA-Lo H2 case

Ratio H2 price to gasoline price (%)	15-year IRR (%)	20-year IRR (%)	30-year IRR (%)
70	10.69	12.64	13.92
80	10.91	12.90	14.23
90	11.12	13.15	14.51
100	11.34	13.40	14.79

The rates of return for the three IEA cases in Hawaii are shown in Table 8.10. The 30-year rate of return on Hawaii is maximized for the IEA-Lo H2 case (14.79 %), while the rate of return in California was highest (12.86 %) also for the IEA-Lo H2 case (see Table 8.7).

The price of hydrogen on Hawaii can also be reduced relative to gasoline as shown in Table 8.11. In this case, if hydrogen is sold at a 30 % discount relative to gasoline per mile driven, even the 15-year rate of return is above 10 %.

The cash flow statement for the IEA Lo-H2 case in Hawaii is shown in Table 8.12 on the next page.

Returns without any carbon tax. The returns on investment are still reasonable even without any carbon tax as shown in Tables 8.13 (Los Angeles) 8.14 (Hawaii). For the Los Angeles case, the IEA-Mix case provides the highest 30-year return at 10.94%.

For the Hawaii case, the maximum 30-year IRR drops from 14.79 % with the carbon tax to 11.32 % without the tax, both for the IEA Lo-H2 case.

The 30-year IRRs are summarized in Table 8.15 for both regions with and without the carbon tax. In all cases, adding the carbon tax increases the returns compared to the same scenario without tax. The only anomaly is that the maximum return for Los Angeles switches from the IEA-Lo-H2 case with the carbon tax to the IEA-Mix case without any tax.

Table 8.12 15-year cash flow for the IEA Lo H2 case in Hawaii with CCL carbon fees and FCEVs on the road according to Fig. 8.2

IEA-Lo-H2 HI	2016	2017	2018	2019	2020	2021	2022
All-sector elect. price (₵/kWh)	29.80	29.86	29.58	29.82	30.38	30.90	31.11
Commercial elect price (₵/kWh)	34.60	34.77	34.55	34.87	35.49	36.12	36.36
Citizens' Climate Lobby tax ($/tonne)	16.54	27.56	38.59	49.61	60.64	71.66	82.69
Carbon tax for coal-ST (₵/kWh)	1.84	3.07	4.30	5.53	6.76	7.99	9.22
Carbon tax for NGCC (₵/kWh)	0.84	1.40	1.96	2.52	3.08	3.63	4.19
Average tax (₵/kWh)	1.34	2.24	3.13	4.02	4.92	5.81	6.71
Carbon tax for IGCC+CCS	0.14	0.24	0.34	0.43	0.53	0.63	0.72
Delta tax advantage for IGCC+CCS	1.20	1.99	2.79	3.59	4.39	5.19	5.98
Capacity price (₵kWh)	0.0		0.00	0.00	0.00	0.00	0.00
Gasoline cost ($/gallon)	2.58	2.64	2.64	2.65	2.68	2.72	2.76
Diesel fuel costs ($/gallon)	3.99	4.50	4.52	4.59	4.65	4.73	4.85
Coal costs ($/MBTU)	4.057	4.178	4.304	4.433	4.566	4.703	4.844
Hydrogen equiv price for FCEVs ($/kg)	6.281	6.442	6.443	6.460	6.535	6.642	6.729
Merchant H2 ($/kg)	9.550	9.550	9.550	9.550	9.550	9.550	9.550
Expenses							
IGCC+CCS+LH2 plant capital cost	$3,368,934,476						
LH2 station capex ($M)	3,470,842	6,941,685	6,941,685	6,941,685	6,941,685	10,412,527	10,412,527
CO_2 transport	13,980,615						
CO_2 storage	71,633,900						
Cumulative capex	$3,458,019,834	$3,464,961,518	$3,471,903,203	$3,478,844,888	$3,485,786,572	$3,496,199,099	$3,506,611,626
Construction financing	$404,272,137						
T&I	$–	$69,299,230	$69,438,064	$69,576,898	$69,715,731	$69,923,982	$70,132,233
Coal MBTU req'd/year			26,983,510	53,967,020	53,967,020	53,967,020	53,967,020
IGCC+CCS O&M		$–	$34,547,297	$69,094,594	$69,094,594	$69,094,594	$69,094,594
Carbon tax			2,997,044.25	16,783,447.81	21,578,718.61	26,373,989.42	31,169,260.22

(continued)

Table 8.12 (continued)

IEA-Lo-H2 HI	2016	2017	2018	2019	2020	2021	2022
Total costs (W/O tax)	$3,862,291,971	$76,240,915	$113,529,810	$384,846,146	$392,161,835	$403,233,089	$411,055,265
Operating cost multiplier	0	0	0.5	1	1	1	1
LH2 annual kWh req'd for LH2 plant	–	–	12,790,787	62,014,932	75,089,271	90,756,512	109,456,346
Annual kWh generated	–		1,463,147,760	2,926,295,520	2,926,295,520	2,926,295,520	2,926,295,520
Net annual kWh delivered	–		1,450,356,973	2,864,280,588	2,851,206,249	2,835,539,008	2,816,839,174
# of FCEVs supported (millions)	0.01786	0.0217	0.0264	0.03196	0.04	0.05	0.06
Hydrogen consumed by FCEVs (kg/year)	0.0000	0	2,192,040	5,316,712	6,438,020	7,781,706	9,385,479
# FCEVs added next year (millions)	0.0038	0.0046	0.0056	0.0067	0.0081	0.0096	0.0115
Number of new stations required	1.0	2.0	2.0	2.0	2.0	3.0	3.0
Total # of LH2 stations	1.0	3	5	7	9	12	15
LH2 transportation cost ($)	–	$–	$2,393,705	$5,807,397	$7,034,208	$8,505,267	$10,263,274
LH2 station electricity cost ($)	–	$–	$32,799	$92,676	$121,287	$164,567	$207,112
LH2 stations non-fuel O&M	–	–	47,587	133,244	171,314	228,419	285,524
Purchased electricity (kWh)	–	–	–	–	–	–	–
Purchased electricity cost			–	–	–	–	–
Total costs (w/o carbon tax)	3,865,762,813	83,182,600	116,003,902	390,879,463	399,488,644	412,131,342	421,811,175
Revenue							
Electricity sales	$–		$234,783,514	$956,929,872	$991,379,102	$1,023,322,312	$1,044,771,138
Hydrogen sales to FCEVs	$–		$7,061,630	$34,344,771	$42,074,739	$51,684,942	$63,152,124
Merchant H2 sales	$–	$–	$9257	$18,513	$18,513	$18,513	$18,513
Total revenue	$–	$–	$241,854,400	$991,293,157	$1,033,472,355	$1,075,025,767	$1,107,941,775
Cash flow	*$(3,862,291,971)*	*$(76,240,915)*	*$125,850,498*	*$600,413,694*	*$633,983,710*	*$662,894,425*	*$686,130,600*
Cumulative cash flow	$(3,862,291,971)		$(3,736,441,473)	$(3,136,027,778)	$(2,502,044,068)	$(1,839,149,643)	$(1,153,019,044)
15-year IRR	11.34%						
20-year IRR	13.40%						

(continued)

Table 8.12 (continued)

IEA-Lo-H2 HI	2016	2017	2018	2019	2020	2021	2022
30-year IRR	14.79%						
Cash flow—HI	$(3,963,800,093)	$(110,043,662)	$13,697,828	$173,734,080	$117,161,080	$13,044,427	$(130,331,309)

IEA-Lo-H2 HI	2023	2024	2025	2026	2027	2028	2029	2030
All-sector elect. price (₵/kWh)	31.37	31.38	31.73	31.85	32.00	32.07	31.95	31.95
Commercial elect price (₵/kWh)	36.65	36.65	37.03	37.13	37.27	37.36	37.18	37.14
Citizens' Climate Lobby tax ($/tonne)	93.71	104.74	115.76	126.79	137.81	148.84	159.86	170.89
Carbon tax for coal-ST (₵/kWh)	10.45	11.68	12.91	14.14	15.37	16.60	17.83	19.06
Carbon tax for NGCC (₵/kWh)	4.75	5.31	5.87	6.43	6.99	7.55	8.11	8.67
Average tax (₵/kWh)	7.60	8.50	9.39	10.29	11.18	12.07	12.97	13.86
Carbon tax for IGCC+CCS	0.82	0.92	1.01	1.11	1.20	1.30	1.40	1.49
Delta tax advantage for IGCC+CCS	6.78	7.58	8.38	9.18	9.97	10.77	11.57	12.37
Capacity price (₵kWh)	0.00	0.00	0.00	0.00	0.00	0.00	0.00	0.00
Gasoline cost ($/gallon)	2.80	2.85	2.89	2.94	2.98	3.03	3.08	3.14
Diesel fuel costs ($/gallon)	4.94	5.03	5.13	5.23	5.33	5.43	5.53	5.63
Coal costs ($/MBTU)	4.989	5.139	5.293	5.452	5.615	5.784	5.957	6.136
Hydrogen equiv price for FCEVs ($/kg)	6.830	6.938	7.047	7.163	7.275	7.396	7.520	7.649
Merchant H2 ($/kg)	9.550	9.550	9.550	9.550	9.550	9.550	9.550	9.550
Expenses								
IGCC+CCS+LH2 plant capital cost								
LH2 station capex ($M)	13,883,369	13,883,369	17,354,212	17,354,212	20,825,054	24,295,896	24,295,896	27,766,739
CO_2 transport								
CO_2 storage								
Cumulative capex	$3,520,494,996	$3,534,378,365	$3,551,732,577	$3,569,086,789	$3,589,911,843	$3,614,207,739	$3,638,503,635	$3,666,270,374

(continued)

Table 8.12 (continued)

IEA-Lo-H2 HI	2023	2024	2025	2026	2027	2028	2029	2030
Construction financing								
T&I	$70,409,900	$70,687,567	$71,034,652	$71,381,736	$71,798,237	$72,284,155	$72,770,073	$73,325,407
Coal MBTU req'd/year	53,967,020	53,967,020	53,967,020	53,967,020	53,967,020	53,967,020	53,967,020	53,967,020
IGCC+CCS O&M	$69,094,594	$69,094,594	$69,094,594	$69,094,594	$69,094,594	$69,094,594	$69,094,594	$69,094,594
Carbon tax	35,964,531.02	40,759,801.82	45,555,072.63	50,350,343.43	55,145,614.23	59,940,885.03	64,736,155.84	69,531,426.64
Total costs (W/O tax)	$422,646,118	$431,001,400	$443,139,268	$452,055,893	$464,769,863	$477,818,049	$487,668,135	$473,572,667
Operating cost multiplier	1	1	1	1	1	1	1	1
LH2 annual kWh req'd for LH2 plant	131,670,093	157,909,771	188,699,490	224,547,446	265,907,429	313,130,096	366,406,434	425,708,729
Annual kWh generated	2,926,295,520	2,926,295,520	2,926,295,520	2,926,295,520	2,926,295,520	2,926,295,520	2,926,295,520	2,926,295,520
Net annual kWh delivered	2,794,625,427	2,768,385,749	2,737,596,030	2,701,748,074	2,660,388,091	2,613,165,424	2,559,889,086	2,500,586,791
# of FCEVs supported (millions)	0.07	0.08	0.10	0.12	0.14	0.16	0.19	0.22
Hydrogen consumed by FCEVs (kg/year)	11,290,619	13,541,040	16,181,690	19,256,155	22,803,354	26,853,360	31,422,553	36,508,557
# FCEVs added next year (millions)	0.0135	0.0159	0.0185	0.0213	0.0243	0.0275	0.0306	0.0335
Number of new stations required	4.0	4.0	5.0	5.0	6.0	7.0	7.0	8.0
Total # of LH2 stations	19	23	28	33	39	46	53	61
LH2 transportation cost ($)	$12,351,202	$14,818,682	$17,715,789	$21,090,446	$24,986,307	$29,437,073	$34,460,530	$40,055,308
LH2 station electricity cost ($)	$264,446	$320,092	$393,667	$465,222	$551,928	$652,506	$748,178	$860,282
LH2 stations non-fuel O&M	361,664	437,803	532,978	628,153	742,362	875,607	1,008,851	1,161,130
Purchased electricity (kWh)	–	–	–	–	–	–	–	–
Purchased electricity cost	–	–	–	–	–	–	–	–
Total costs (w/o carbon tax)	435,623,430	446,577,977	461,781,702	474,239,713	491,050,461	508,783,235	523,885,695	515,649,388
Revenue								
Electricity sales	$1,066,266,752	$1,078,691,931	$1,098,058,636	$1,108,576,012	$1,116,638,488	$1,119,456,690	$1,114,019,514	$1,108,152,635

(continued)

Table 8.12 (continued)

IEA-Lo-H2 HI	2023	2024	2025	2026	2027	2028	2029	2030
Hydrogen sales to FCEVs	$77,109,411	$93,941,553	$114,032,182	$137,926,678	$165,897,259	$198,595,159	$236,310,570	$279,241,710
Merchant H2 sales	$18,513	$18,513	$18,513	$18,513	$18,513	$18,513	$18,513	$18,513
Total revenue	$1,143,394,676	$1,172,651,997	$1,212,109,331	$1,246,521,203	$1,282,554,261	$1,318,070,362	$1,350,348,598	$1,387,412,858
Cash flow	*$707,771,247*	*$726,074,020*	*$750,327,629*	*$772,281,490*	*$791,503,800*	*$809,287,127*	*$826,462,903*	*$871,763,471*
Cumulative cash flow	$(445,247,797)	$280,826,223	$1,031,153,852	$1,803,435,342	$2,594,939,142	$3,404,226,269	$4,230,689,172	$5,102,452,643
15-year IRR								
20-year IRR								
30-year IRR								
Cash flow—HI	$1,020,338,786	$1,040,161,466	$1,058,952,082	$1,078,044,431	$1,098,214,590	$1,125,840,952	$1,151,727,708	$1,176,634,548

Table 8.13 Returns on investment for Los Angeles without any carbon tax

	20-year IRR (%)	30-year IRR (%)
IEA-Mix	8.69	10.94
IEA-Lo H2	8.18	10.40
IEA-Hi H2	8.21	10.69

Table 8.14 Returns on investment for the Hawaii case without any carbon tax

	20-year IRR (%)	30-year IRR (%)
IEA-Mix	6.22	8.88
IEA-Lo H2	9.35	11.32
IEA-Hi H2	−0.36	4.32

Table 8.15 30-year rates of return for the three IEA scenarios for Los Angeles and Hawaii with and without a carbon tax

	Los Angeles		Hawaii	
	No tax (%)	With tax (%)	No tax (%)	With tax (%)
IEA-Mix	10.94	12.32	8.88	12.13
IEA-Lo H2	10.40	12.68	11.32	14.79
IEA-Hi H2	10.69	11.08	4.32	7.05

References

1. IEA Greenhouse Gas R&D Programme (IEA GHG) (2008) Co-production of hydrogen and electricity by coal gasification: updated economic analysis. Report # 2008/9, August 2008. Available at: http://www.climit.no/en/Protected%20Library/IEA%20reports/2008_09_CO_production_of_hydrogen_and_electricity_by_coal_gasification_with_CO2_capture_updated_economic_analysis.pdf
2. The IHS North American Power Capital Cost Index (PCCI). Available at: https://www.ihs.com/info/cera/ihsindexes/
3. Yang C, Ogden J Determining the lowest-cost hydrogen delivery mode. Institute for Transportation Studies, University of California at Davis, report UCD-ITS-RP-07-40. Available at: https://its.ucdavis.edu/research/publications/?frame=https%3A%2F%2Fitspubs.ucdavis.edu%2Findex.php%2Fresearch%2Fpublications%2Fpublication-detail%2F%3Fpub_id%3D1162
4. Halvorson T et al (1996) Hydrogen production and fueling system infrastructure for PEM fuel cell-powered vehicles. Praxair, Inc. for the Ford Motor Company, 12 April 1996
5. Moore R (1996) Ford hydrogen infrastructure study, summary report. Air Products and Chemicals Inc., March 1996
6. Ramsden T et al (2013) Hydrogen pathways: updated cost, well-to wheels energy use, and emissions for current technology stations of ten hydrogen production, delivery and distribution scenarios. The National Renewable Energy Laboratory (NREL/TP-6A10-60528, March 2013. Available at: http://www.nrel.gov/docs/fy14osti/60528.pdf
7. Kelly B (2007) Liquefaction and pipeline costs. Nexant, Inc., May 2007. Available at: http://www1.eere.energy.gov/hydrogenandfuelcells/pdfs/deliv_analysis_kelly_lh2.pdf
8. Los Angeles Almanac, Available at: http://www.laalmanac.com/transport/tr02.htm

9. Davis S et al Transportation energy data book: edition 34. Oak Ridge National Laboratory, Oak Ridge, Tennessee, ORNL-6991. Available at: http://cta.ornl.gov/data/index.shtml
10. Yang C, Ogden J (2007) Hydrogen delivery and refueling infrastructure; Advanced energy pathways (AEP) project, Task 4.1 technology assessments of vehicle fuels and technologies, public interest energy research program. Prepared by the Institute of Transportation Studies at UC-Davis for the California Energy Commission, June 2007
11. Fesharaki F et al (2003) Final Report on Hawaii hydrocarbon outlook. FACTS, Inc. for the Hawaii Energy Policy Project, University of Hawaii at Manoa, Jan 2003. Available at: http://www.hawaiienergypolicy.hawaii.edu/working-groups/hydrocarbon-future-energy-security/_downloads/hydrocarbons-outlook.pdf
12. BP Statistical Review of World Energy, June 2015. Available at: http://www.bp.com/statisticalreview

Chapter 9
Solar-Hydrogen Generation Systems

Abstract While we are advocating the use of coal with CO_2 sequestration for most of the 21st century to provide low carbon electricity and hydrogen to power fuel cell electric vehicles, coal is a finite resource and eventually society will have to generate electricity and hydrogen from low- or zero-carbon sources such as wind or solar (or nuclear). In this chapter we explore the option of building solar photovoltaic (PV) generators with the hydrogen produced during sunlight hours stored to be used to generate electricity when the sun is not shining. We show that the lowest cost and lowest greenhouse gas (GHG) current alternative electricity generators available today [Natural gas combined cycle (NGCC) generators] to back up intermittent renewables would not allow us to meet our greenhouse gas goal of cutting GHGs to 80% below 1990 levels by 2050. In fact, even if *all* U.S. electricity was generated from PV solar with NGCC backup, then the GHG emissions would still be *four times* the GHG emissions necessary to achieve the societal GHG goal. We also show that an organization building these PV-Hydrogen systems could earn 30-year internal rates of return above 9% without a carbon tax, and between 11 and 12% with the Citizen's Climate Lobby carbon fee and dividend.

A PV[1] solar-hydrogen system would include the following components as shown in Fig. 9.1

- A PV array to generate electricity from the sun
- An electrolyzer to convert some electricity to hydrogen
- A compressor to pressurize the hydrogen for storage
- A storage system
- A hydrogen generator such as a fuel cell to regenerate electricity from the stored hydrogen when the sun is not shining or when the solar output cannot meet the load.

---

[1]While this chapter explores the economics of using PV systems to provide zero GHG electricity and fuel for FCEVs, wind energy could also provide zero GHG energy for a similar hydrogen storage system; we did not analyze wind energy here.

C.E. Thomas, *Stopping Climate Change: the Case for Hydrogen and Coal*,
Lecture Notes in Energy 35, DOI 10.1007/978-3-319-31655-0_9

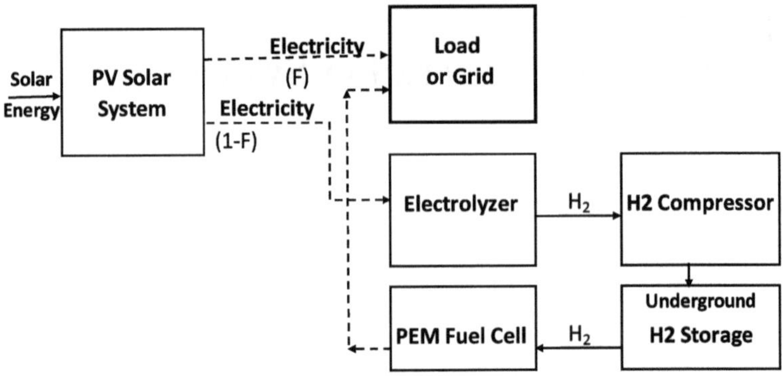

Fig. 9.1 Block diagram of a PV-hydrogen generation system

The PV system is sized to provide 300 kW *average* power every hour of the year. We assume that PV solar energy when the sun is shining is all fed directly to the load to avoid losses in the electrolyzer-storage-H2 generator system (F = 1); if the solar energy is not sufficient to meet the load, then the stationary fuel cell provides the deficit using stored hydrogen; if the solar energy exceeds the load energy required, then the excess solar energy goes to the electrolyzer (minus enough electricity to run the hydrogen compressor) to make hydrogen for later use. After the sun sets and on cloudy days, stored hydrogen is fed to the stationary PEM fuel cell generator to supply the average 300 kW of power to the load at all hours, day and night.

9.1 PV System Sizing

We sized this system to generate an average of 300 kW of electrical power day and night.[2] The PV solar arrays will have to supply more than 300 kW peak power to generate enough electricity to power the load at night and on cloudy days and when the solar energy is not sufficient to power the load. To estimate the peak PV array power required, we analyzed solar energy available for Los Angeles as summarized in Table 9.1, using the average measured solar radiation taken from the National Renewable Energy Laboratory (NREL) "PVWatts®" data based on 19 year's of solar data (1961–1990) from the Los Angeles airport [1]. These PVWatts data include 14% of losses accounting for mismatch losses, wiring and connector losses and light-induced degradation of 1.5% and inverter efficiency of 96% [2]. We also

[2]As shown in Table 7.3 in Chap. 7, the average U.S. "Big Box" store consumes approximately 335 kW of electricity, so a 300-kW system capacity would be suitable for an average U.S. store or warehouse.

use their estimate of 15% efficiency for the solar modules and 96% efficiency for the DC to AC inverter. We compared two systems:

- A fixed-tilt flat-plate PV system tilted at 34°, the latitude of Los Angeles, and
- A single-axis (East-West) tracking system in Los Angeles also tilted at 34° above the horizontal in the North-South plane.

Table 9.1 illustrates the energy flows in a fixed-tilt PV system in Los Angeles, using the average monthly solar irradiance for Los Angeles and the system component efficiencies listed in Table 9.2. For this fixed-tilt PV system, energy is withdrawn from storage from November through January, while energy is delivered to storage from February through October in Los Angeles. The computer model selects the PV area and power that will exactly match December and January hydrogen storage levels. As shown in Table 9.1, the hydrogen storage reaches a maximum of 201,215 kWh at the end of October; however, we increase storage by 15% to account for months when the solar irradiance is above the 15-year average used in Table 9.1. Thus we assume a hydrogen storage system of 231,398 kWh instead of the 201,215 shown above. This 15% larger storage volume sets the cost of the geologic storage system.

Table 9.1 is for a fixed tilt PV system. Table 9.3 shows similar data for a one-axis tracking system in Los Angeles. Since the tracking system collects more of the solar radiation, the peak hydrogen storage grows from 201,215 kWh for a fixed-tilt PV system to 224,184 kwh for the tracking system, both at the end of October. As with the fixed-tilt system, we add a 15% cushion to the tracking system storage capacity, building a hydrogen storage system with a capacity of 257,812 kWh to account for years with higher than average solar irradiance.

9.2 PV System Costs

A joint Solar Energy Industries Association (SEIA) and GTM Research report stated that the price of utility PV systems in the fourth quarter of 2015 had fallen to $1.33/W for fixed systems and to $1.54/W for single-axis tracking PV systems [3]. In addition, GTM Research has projected that PV module prices will most likely decrease to only 36 ₵/Watt by 2017 [4], compared to PV module costs of 60 ₵/W used in the fourth quarter 2105 PV system estimates listed above. We estimate that a PV module price of 36 ₵/Watt by 2017 will reduce the PV system prices to approximately $1.12 per Watt for fixed arrays and to $1.33 per watt for single-axis tracking systems without any further reductions in other subsystems. These are PV system *prices* including all subsystems such as inverters, array mounts and all other balance of system (BOS) components plus installation labor, overhead, profits and taxes. We use the lower future costs in this model, since we are assuming that coal-based IGCC + CCS systems provide the bulk of the electricity and hydrogen for FCEVs and MHEs throughout most of the 21st century. So using the 2015–

Table 9.1 Estimated average energy flows for a fixed-tilt PV system in Los Angeles with a tilt angle of 34° (the latitude of Los Angeles)

	Solar irradiance (kWh/m2/day)		Solar energy to array	Electricity to load	Electricity from fuel cell	Energy from storage	Energy to storage	Net energy in storage end of month
	Fixed array (34° tilt)	One-axis tracking	kWh/mon.	kWh/mon.	kWh/mon.	kWh/mon.	kWh/mon.	kWh
Jan	4.15	4.77	1,431,920	177,329	45,871	79,887	0	42,581
Feb	4.59	5.29	1,430,473	177,150	24,450	42,581	0	0
Mar	5.62	6.69	1,939,130	240,142	0	0	10,677	10,677
Apr	6.17	7.3	2,060,229	255,139	0	0	24,665	35,342
May	5.98	7.21	2,063,345	255,525	0	0	20,371	55,713
Jun	6.45	7.7	2,153,724	266,717	0	0	31,962	87,675
Jul	6.82	8.21	2,353,180	291,418	0	0	42,991	130,666
Aug	6.75	8.23	2,329,027	288,427	0	0	41,106	171,772
Sep	6.14	7.31	2,050,211	253,898	0	0	23,883	195,655
Oct	5.43	6.34	1,873,573	232,023	0	0	5560	201,215
Nov	4.95	5.76	1,652,858	204,690	11,310	19,697	0	181,518
Dec	4.43	5.09	1,528,532	189,293	33,907	59,050	0	122,468

Table 9.2 Estimated efficiencies and capital costs of system components

	Efficiencies (%)	Capital costs
PV	22.8	$1301/kW
Inverter	94.0	(included above)
Electrolyzer	68.5	$ 619/kW
Compressor	92.0	
Fuel cell	58.0	$1812/kW
Storage	99.0	$1.19/kWh

2017 PV price estimates is conservative, since further PV system cost reductions are likely by the time society begins running out of domestic coal.

9.3 PEM Fuel Cell Costs

We have cited the expected costs of PEM fuel cell systems by James et al. from Strategic Analysis Inc. (SAI) in Chap. 7. James estimates a cost of $2011/kWe for a 500-kWe fuel cell system, based on a production quantity of ten units, which we take to represent "current costs." By the time we implement PV-hydrogen systems several decades from now, presumably many units will be built and there will be other reductions in cost as a result of further furl cell system development. For the longer term, we assume that 100 PEM fuel systems are built each year, which will drive down the costs even lower. We estimate a cost of $1812/kWe for 300 kWe PEM fuel cells system in production quantities of 100 units per year, dropping to $1732/kWe for a 1-MW fuel cell system based on scaling data from James et al.

9.4 Electrolyzer System Costs

We average Electrolyzer costs from two sources: Colella et al. from SAI [5] and Ainscough et al. from NREL [6] for Proton Exchange Membrane (PEM) electrolyzers as shown in Table 9.4. We estimate both current costs and future costs and use the future costs since it will be many decades before society runs low on the coal proposed here to provide both electricity and hydrogen for FCEVs. We assume that the electrolyzers must be replaced every 15 years to start and every 26 years in the future with a 70% salvage value.

Table 9.3 Hydrogen storage values for a one-axis tracking system in Los Angeles

	Solar irradiance (kWh/m2/day)		Solar energy to array	Electricity to load	Electricity from fuel cell	Energy from storage	Energy to storage	Net energy in storage end of month
	Fixed array (34° tilt)	One-axis tracking	kWh/mon.	kWh/mon.	kWh/mon.	kWh/mon.	kWh/mon.	kWh
Jan	4.15	4.77	1,401,172	173,521	49,679	86,518	0	48,390
Feb	4.59	5.29	1,403,541	173,815	27,785	48,390	0	0
Mar	5.62	6.69	1,965,166	243,366	0	0	12,709	12,709
Apr	6.17	7.3	2,075,179	256,990	0	0	25,832	38,541
May	5.98	7.21	2,117,914	262,283	0	0	24,630	63,171
Jun	6.45	7.7	2,188,887	271,072	0	0	34,706	97,877
Jul	6.82	8.21	2,411,661	298,660	0	0	47,555	145,432
Aug	6.75	8.23	2,417,536	299,388	0	0	48,013	193,445
Sep	6.14	7.31	2,078,022	257,342	0	0	26,054	219,499
Oct	5.43	6.34	1,862,355	230,634	0	0	4685	224,184
Nov	4.95	5.76	1,637,402	202,776	13,224	23,031	0	201,153
Dec	4.43	5.09	1,495,171	185,162	38,038	66,245	0	134,908

Table 9.4 Estimated PEM electrolyzer costs including power electronics

	Current	Future
Colella	$1400.7/kW	$670.4/kW
Ainscough	$1205.2/kW	$567.9/kW
Average	$1303.0/kW	$619.2/kW

Table 9.5 Estimated annual operation and maintenance (O&M) costs for the PV-hydrogen system (LA tracking)

	Basis:	Annual O&M cost
PV O&M ($/kW-yr)	24.69	$ 54,394
Electrolyzer O&M (% of capex)	5.0%	$ 68,202
Storage O&M (% of capex)	0.5%	$ 1733
Compressor O&M	6.0%	$ 8708
PEM FC O&M ($/kW-yr)	20	$ 4000
Total annual O&M		$ 137,038

9.5 Operation and Maintenance (O&M) Costs

Our estimate of the PV/Hydrogen system annual O&M costs are summarized in Table 9.5. In addition, we add 1.75% of the capital expenses each year to cover taxes and insurance on the equipment. A 12% finance charge is added the first year to all capital expenses.

9.6 Geologic Storage Costs

The cost of storing hydrogen in underground caverns is relatively low. Lord et al. from Sandia National Laboratory estimate a cost of a very large capacity (1.912 million kg) underground hydrogen storage system (Table 2 of reference [7]). For comparison, the storage required for the PV-hydrogen systems here is a maximum of 6222 kg, or 300 times smaller than the Sandia system. These cost estimates include all mining, piping, wells and compressor costs. To be conservative, we use the higher (hard rock) storage cost; storage costs will be less in areas with access to salt caverns, aquifers or depleted oil or gas reservoirs. To scale down the Sandia storage system to the smaller storage required here, we consider the compressors separately from all the other components such as piping, mining, wells, etc. Elgowainy et al. at the Argonne National Laboratory have shown hydrogen compressor cost estimates that vary as the 0.61 power of the hydrogen flow rate [8]. The Sandia storage system is based on hydrogen flow rates as high as 4920 kg/hr, while the highest flow rate for the PV-H2 system is only 45.5 kg/hr for our assumed 300 kWe load. As shown in Table 9.6, the compressor cost would

Table 9.6 Hydrogen compressor cost scaling estimates

Compressor scaling	
H2 req'd (kWh)	517.241379
H2 req'd (kg/hr)	15.489979
Sandia H2 output (kg/hr)	4920
ANL power law scaling factor	0.61
Cost ratio	0.02977
Sandia compressor cost ($M)	$ 27.539
PV-H2 compressor cost ($M)	$ 0.820

Table 9.7 Estimated PV-hydrogen storage costs for a hard rock storage system based on scaling from the much larger Sandia underground hydrogen storage system

For hard rock storage	Sandia	PV-H2	Cost ratio
Stored hydrogen (kg)	1,912,000	6222	0.003254
Total capex ($M)	$ 53.11		
Compressor cost ($M)	$ 27.54	$ 0.820	
Non-compressor costs ($M)	$ 25.57	$ 0.083	
Total capex ($M)	$ 53.11	$ 0.903	

decrease from $27.5 million for the Sandia system down to $820,000 for the PV-H2 compressor system using the 0.61 power scaling factor.

We assume that all other components of the storage system scale with the maximum hydrogen stored in the system as shown in Table 9.7. The total storage system costs decrease from $53.1 million for the Sandia system to $903,000 for the PV-hydrogen system with single-axis tracking.[3]

9.7 Carbon Tax

We assume that a carbon tax or fee is assessed on electricity based on the Citizen's Climate Lobby proposal described in Appendix A. This plan assesses a fee of $15/ton ($16.54/metric tonne) beginning in 2016,[4] and increasing by $10/ton ($11.03/tonne each year), with all the revenue returned to tax-payers in the form of a dividend. The utility must include this tax or fee in their bids into the wholesale electricity markets in the US. The tax or fee is then passed on to the consumer (the "Big Box" store or warehouse owner in this case). The owner of the PV/hydrogen system with low greenhouse gas emissions will avoid paying most of this carbon tax from fossil fuel generators, which effectively becomes an avoided cost "revenue" to the facility owner.

[3]The storage system is slightly lower for the fixed tilt PV system at $892,000 due to the reduced storage capacity required.

[4]Or the first year of the PV-hydrogen system operation.

Table 9.8 Total capital costs for a 300-kWe PV-hydrogen system in Los Angeles

	Fixed array	One-axis tracking
PV peak power (kW)	377.763	402.403
Storage req'd (kWh)	180,335	207,778
PV cost	$ 509,633	$ 523,571
Electrolyzer cost	$ 242,519	$ 249,151
compressor cost	$ 168,186	$ 183,528
Geo storage cost	$ 892,174	$ 903,165
FC Cost	$ 543,654	$ 543,654
Total Capex	$ 2,356,166	$ 2,403,069

Table 9.9 Rates of return and capital expenses for a PV-hydrogen system in Los Angeles providing continuous 300 kWe electricity

	No carbon tax		With CCL carbon tax	
	Fixed array	One-axis tracking	Fixed array	One-axis tracking
15-yrIRR	7.14%	6.69%	10.49%	10.04%
20-yr IRR	9.04%	8.64%	12.25%	11.84%
30-yr IRR	10.22%	9.86%	13.14%	12.76%
Capex ($M)	$ 2.356	$ 2.403	$ 2.356	$ 2.403

9.8 Total PV/Hydrogen System Cost

The total PV/Hydrogen storage system cost of just over $2.4 million is summarized in Table 9.8 for a Los Angeles system with a peak PV power of 380–402 kW and an average power of 300 kWe to the load. This sum includes $168K–$183 K for a hydrogen compressor at the PV facility to pressurize the hydrogen pipeline to the storage site (The storage system also includes its own hydrogen compressor as described above to fill the underground storage wells.) The fixed tilt PV system is slightly less expensive, due primarily to the reduced hydrogen storage requirement and to a lesser extent the lower PV power required.[5]

9.9 Summary of Rates of Return

The capital costs and internal rates of return for PV-hydrogen systems are summarized in Table 9.9. The fixed tilt array has marginally higher IRRs due to lower capital costs.

[5]The one-axis tracking system produces slightly more hydrogen, which requires a larger hydrogen storage system.

Table 9.10 Cash-flow statement for a fixed-tilt PV/hydrogen system in Los Angeles providing 300-kWe electricity 24/7 with the CCL carbon tax

	2016	2017	2018	2019	2020	2021	2022	2023	2024	2025	2026	2027	2028	2029	2030	2031
Average Power (kW)	300.0															
Peak PV Power (kW)	391.69															
PV cost (includes inverter & BOS)	$1,301.11/kW															
PV Capex	$ 509,632.93															
Electrolyzer Capex	$ 242,518.68															
Compressor Capex	$ 168,185.71															
Geologic Storage Capex	$ 892,174.06															
PEM FC Capex	$ 543,654.43															
Total Capex	$ 2,356,165.81															
Finance Charge:12 %	$282,739.90															
Investment tax credit	$0.00															
SMR+PSA	$0.00															
Cascade storage	$0.00															
Refrigeration	$0.00															
700-bar compressor	$0.00															
Electrical	$0.00															
Dispenser	$0.00															
Controls, Safety & "other" capex	$0.00															
Total H2 station costs	$0.00															
Annual capaciry Factor	95%															
Gasoline cost ($/gallon)	2.02	2.20	2.37	2.60	2.74	2.82	2.86	2.91	2.94	2.97	3.03	3.08	3.11	3.16	3.19	3.25
Electricity per year (kWh/yr)	2,496,600	2,496,600	2,496,600	2,496,600	2,496,600	2,496,600	2,496,600	2,496,600	2,496,600	2,496,600	2,496,600	2,496,600	2,496,600	2,496,600	2,496,600	2,496,600
T&I		$41,233	$41,233	$41,233	$41,233	$41,233	$41,233	$41,233	$41,233	$41,233	$41,233	$41,233	$41,233	$41,233	$41,233	$41,233
O&M		$42,348.79	$42,349	$42,349	$42,349	$42,349	$42,348.79	$42,349	$42,349	$42,348.79	$42,348.79	$42,348.79	$42,348.79	$42,348.79	$42,348.79	$42,348.79
PEM FC & Electrolyzer replacment costs														$ -	$ -	$ -
Total annual costs	$ 2,638,905.70	$83,582	$83,582	$83,582	$83,582	$83,582	$83,582	$83,582	$83,582	$83,581.69	$83,581.69	$83,581.69	$83,581.69	$83,581.69	$83,581.69	$83,581.69
	2016	2017	2018	2019	2020	2021	2022	2023	2024	2025	2026	2027	2028	2029	2030	2031
Com. electricity price (c/kWh)	$15.54	$15.56	$15.51	$15.53	$15.49	$15.35	$15.16	$14.96	$14.80	$14.79	$14.73	$14.60	$14.50	$14.37	$14.27	$14.22
Avoided Electr. Carbon tax	$ 16,410.9	$ 26,702.3	$ 36,943.8	$ 46,174.5	$ 48,795.9	$ 54,607.6	$ 62,458.1	$ 70,736.0	$ 78,848.1	$ 78,363.6	$ 86,434.4	$ 94,295.4	$ 97,882.7	$ 104,733.4	$ 109,655.4	$ 113,777.8
Carbon tax ($/tonne)	$16.54	$27.56	$38.59	$49.61	$60.64	$71.66	$82.69	$93.71	$104.74	$115.76	$126.79	$137.81	$148.84	$159.86	$170.89	$181.91
Electricity avoided costs		$388,563.86	$387,309.19	$387,758.95	$386,709.96	$383,177.39	$378,436.20	$373,535.77	$369,560.76	$369,219.90	$367,751.73	$364,379.79	$361,947.63	$358,864.78	$356,233.99	$355,056.82
Total Revenue	$ -	$415,266.16	$424,253.00	$433,933.50	$435,505.89	$437,784.98	$440,894.31	$444,271.77	$448,408.82	$447,583.54	$454,186.15	$458,675.14	$459,830.38	$463,598.23	$465,889.39	$468,834.62
cash flow	(2,638,906)	331,684	340,671	350,352	351,924	354,203	357,313	360,690	364,827	364,002	370,604	375,093	376,249	380,017	382,308	385,253
Cumulative cash flow	(2,638,906)	(2,307,221)	(1,966,550)	(1,616,198)	(1,264,274)	(910,071)	(552,758)	(192,068)	172,759	536,761	907,366	1,282,459	1,658,708	2,038,724	2,421,032	2,806,285
15-yr IRR	10.49%															
20-yr IRR	12.2503%															
30-yr IRR	13.1371%															

9.10 Cash Flow Statement

The cash flow statement for a Los Angeles PV fixed tilt system is shown in Table 9.10.

9.11 Alternative PV Systems

The previous sections all consider the use of hydrogen to store PV energy for use when the sun is not shining. The PV/Hydrogen system **would eliminate all greenhouse gas emissions** from the power generation sector. In the next two sections, we consider other electricity generators to accommodate intermittent renewable energy sources such as PV solar systems. In Sect. 9.11 we consider the use of natural gas combined cycle (NGCC) power plants to supply electricity when the sun is not shining, and in Sect. 9.11 we analyze the use of nuclear power plants to backup intermittent renewable energy.

9.11.1 PV/NGCC Systems

The greenhouse gas emissions from the NGCC plant are estimated at 507.2 g of CO_2-equivalent/kWh of electricity (see Table 6.2). Averaged over the seasons, the NGCC plant would generate approximately 2.48 million kWh/year, which would produce approximately 1.26 million kg/year of CO_2-equivalent GHG emissions. If we replaced all existing 712 GW of fossil fuel plants[6] in the U.S. with PV solar systems backed up by NGCC plants, then these NGCC plants would generate approximately 1705 million metric tonnes of CO_2-equivalent per year, which is *more than two times* our GHG goal of 784 million metric tonnes from the electricity sector to achieve our over-all societal goal of cutting GHGs by 80% below 1990 levels.

9.11.2 PV/Nuclear Systems

GHG emissions attributed to nuclear plants are very low at 14 g of CO_2-equivalent per kWh of power. So nuclear power plants backing up PV systems would generate only 52 million tonnes of CO2 per year, which is only 6.6% of our target for electricity of 784 million tonnes/year.

[6]As of 2015.

Table 9.11 Key attributes of the three PV backup systems (for a PV system in Los Angeles)

	PV-hydrogen	PV-NGCC	PV-nuclear
Capex ($M)	$2.48	$0.894	$2.251
GHGs (M tonnes/year)	0	3163.47	87.32
GHG goal	784	784	784
GHG ratio to goal	0	4.04	11%

The key attributes of the three types of backup to PV solar systems are summarized in Table 9.11. Using NGCC to backup PV solar is the least costly, but would preclude our achieving our goal of cutting GHGs below 784 million tonnes per year. The PV-hydrogen system has zero GHG emissions but the highest cost. A PV-nuclear system also has high cost but would allow us to meet the GHG reduction goal. Both nuclear and hydrogen PV systems would generate much less than our 784 million tonne GHG goal, which would provide a cushion for other sectors of the economy that might not be able to reach their GHG reduction goals. The PV-hydrogen system has the added advantage of being able to capture most of the solar energy to use at night and during cloudy days, whereas this energy is wasted without energy storage on the other two backup systems. Public acceptance of nuclear power may be difficult given the Fukashima nuclear disaster in Japan, combined with the concern over nuclear weapons proliferation and radioactive waste disposal.

Another advantage of the PV/hydrogen system is its low power that can support widely distributed electricity generation, cutting down the need for vary large, central power plants. A distributed PV/hydrogen system will also reduce electricity transmission costs (not included in this model) and add to the resiliency of the electrical grid in the face of external disruptions due to either weather or terrorist attacks.

References

1. NREL's PVWatts® Calculator. Available at http://pvwatts.nrel.gov/
2. Dobos A (2014) PVWatts user's version 5 manual. The National Renewable Energy Laboratory, Sept 2014. Available at: http://www.nrel.gov/docs/fy14osti/62641.pdf
3. The Solar Energy Industry Association (SEIA) and GTM Research (2016) U.S. Solar Market Insight-2015 Year in Review, reported by the PV Magazine US solar system prices fell 17% in 2015, finds GTM, 16 March 2016. Available at: http://www.pv-magazine.com/news/details/beitrag/us-solar-system-prices-fell-17-in-2015–finds-gtm_100023743/#axzz46UoahGnU
4. Rinaldi N Solar PV module prices to fall to 36 cents per watt by 2017. Greentech Media, Available at: http://www.greentechmedia.com/articles/read/solar-pv-module-costs-to-fall-to-36-cents-per-watt
5. Colella W et al (2014) Techno-economic analysis of PEM electrolysis for hydrogen production. Electrolytic hydrogen production workshop, Golden, Colorado, 27 Feb 2014. Available at: http://energy.gov/sites/prod/files/2014/08/f18/fcto_2014_electrolytic_h2_wkshp_colella1.pdf

6. Ainscough C et al (2014) Hydrogen production cost from PEM electrolysis. DOE Hydrogen and fuel cells program record, 1 July 2014. Available at:https://www.hydrogen.energy.gov/pdfs/14004_h2_production_cost_pem_electrolysis.pdf
7. Lord A et al (2014) Sandia National Laboratory "Geologic storage of hydrogen: scaling up to meet city transportation needs." Int J Hydrogen Energy 39, 15570–15582. Available at: http://www.sciencedirect.com/science/article/pii/S0360319914021223
8. Elgowainy A et al (2012) Hydrogen delivery infrastructure analysis. The DOE 2012 hydrogen program review. The Argonne National Laboratory

Chapter 10
Water Consumption

Abstract Generating electricity places stresses on our planet's limited supply of fossil fuels while also creating pollution in the form of greenhouse gases and local air pollution, but many readers may not realize that current electrical generation plants also consume significant quantities of water that is placing stress on global freshwater resources. We have shown in previous chapters that producing electricity by integrated gasification combined cycle (IGCC) coal plants with carbon capture and storage (CCS) will reduce GHG emissions. In this chapter we show that coal-powered IGCC plants with CCS would also require approximately 39 % less water than the current U.S. mix of electricity generation technologies. We also show that the water required to produce hydrogen is much 20–25 times less than the water required to produce gasoline.

10.1 Global Water Resources

The United Nations estimates that at least 768 million and possibly as many as 3.5 billion people globally do not have access to clean water; that 20 % of the world's aquifers are being "over-exploited (and some critically so)"; that 40 % of the global population is expected to be in areas of "sever water stress" through 2050; that deterioration of global wetlands is reducing the ability of natural ecosystems to purify water; that global demand for water will grow by 55 % by 2050. The UN further estimates that approximately 90 % of all electricity generated around the world is "water intensive"; that cooling of electricity generators in Europe accounts for 43 % of freshwater withdrawals and 50 % in the U.S. [1] Therefore minimizing the use of water in the process of generating electricity is a global imperative.

10.2 Water Requirements for Electricity Production

10.2.1 Water Withdrawals Versus Water Consumption

Most water withdrawn for thermoelectric electricity power generation is used for cooling. Water has four times the heat transfer capacity as air, so most plants use

C.E. Thomas, *Stopping Climate Change: the Case for Hydrogen and Coal*, Lecture Notes in Energy 35, DOI 10.1007/978-3-319-31655-0_10

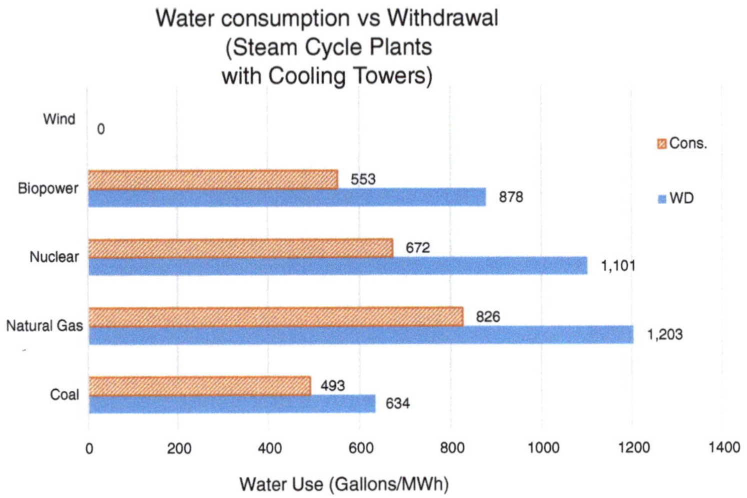

Fig. 10.1 Comparison of water consumption and water withdrawals (WD) from simple cycle thermoelectric power plants using water cooling towers

water to remove waste heat. Figure 10.1 shows the water requirements for various types of simple cycle power plants with cooling towers to cool the water from the steam turbines; two types of water use are shown: water withdrawals and water consumption [2]. Water consumption is less than water withdrawals. Thus water is *withdrawn* from a lake or river, but some is also returned, with the difference being the amount of water consumed in the process, usually by evaporation. While we are most concerned about the total *consumption* of water,[1] the water withdrawals may also be important for local water conditions. Returning warm water to the lake or river may have a deleterious effect on the water ecosystem.

10.2.2 Water Use Versus Generator Type

As power plants become more efficient there will be less waste heat, and hence a reduced demand for cooling water. For example, Fig. 10.2 shows the reduced water requirements for higher efficiency combined cycle natural gas and coal power plants.[2] For the coal plant, water consumption decreases from 493 gallons withdrawn to 380 gallons of water consumed per MWh of electricity generated in a

[1]Of course the evaporated water is not lost, but returns as precipitation at a later time; however the precipitation will not usually occur in the areas that lost the water, and often not in the areas most in need of water.

[2]The coal plant water use in Fig. 8.2 is for an integrated gasification combined cycle (IGCC) power plant.

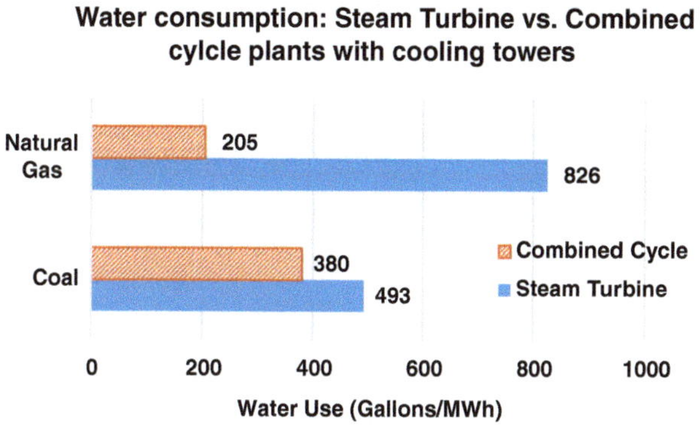

Fig. 10.2 Water consumption for combined cycle natural gas and coal plants using cooling towers

combined cycle plant, a 23 % decrease, while the natural gas water consumption drops from 826 to 205 gallons consumed per MWh of electricity, a 75 % reduction in water consumption [3].

10.2.3 Water Use Versus Cooling Method

Water use also depends on the type of cooling system used at the power plant. The two previous figures were based on water cooling towers. Table 10.1 compares water use for cooling towers, once-through cooling and pond cooling for nuclear and natural gas combined cycle plants. Once-through cooling minimizes the water consumed, but at the expense of much larger water withdrawals. The water withdrawn with once-through cooling is 165 times larger than the water consumed for a nuclear plant, and 114 times larger for an NGCC plant.

10.2.4 Water Use with and Without Carbon Capture and Storage

Carbon capture and storage/sequestration-(CCS) unfortunately decreases the efficiency of power plants, since energy is required to separate out the CO_2 from the gas stream and to compress the CO_2 so it can be transported by pipeline to an underground cavern. This decreased efficiency with CCS therefore increases water use as shown in Fig. 10.3. CCS increases water consumption by 44 % for a coal

Table 10.1 Impact of cooling options on water use for nuclear and natural gas combined cycle plants in gallons of water per MWh of electricity produced. (WD = water withdrawal)

	Nuclear			NGCC		
Cooling options:	WD	Cons.	Ratio	WD	Cons.	Ratio
Tower	1101	672	1.64	255	205	1.24
Once-through	44,350	269	164.9	11,380	100	113.8
Pond	7050	610	11.6	5950	240	24.8

IGCC plant and by 92 % for a NGCC plant.From previous chapters coal-based IGCC plants offer our best option for reducing GHGs while eliminating our dependence on Middle East fossil fuel supplies. While a coal-powered IGCC plant with CCS does consume more water than a Natural gas combined cycle plant (549 vs. 393 gallons/MWh), this is still less than the 826 gallons/MWh of water consumed by single-cycle natural gas steam plants today without CCS.

10.3 Water Requirements for Hydrogen Production

Water is also consumed in the process of making hydrogen. As shown in Table 10.2, making hydrogen at a coal-based IGCC plant with carbon capture and storage (CCS) has a relatively low water requirement of just under 3 gallons of water per kg of hydrogen [4]. Making hydrogen from biomass or from natural gas by distributed steam methane reforming (SMR) requires even less water at the hydrogen production plant, but much more water is required to grow the biomass plants. Some readers might be surprised to see that wind energy used to electrolyze water requires over 100 times more water than making hydrogen from coal at an IGCC plant according to the National Renewable Energy Laboratory (NREL). Wind electricity does not require any water for cooling of the wind turbines, but some electrolyzers[3] must be cooled, accounting for most of the water consumed in the wind electrolysis case.

For the biomass case, we must also include the water used to produce the biomass in addition to the water consumed in generating hydrogen from that biomass. As shown in Table 10.3, the average water consumption for all US crops is approximately 61 m^3 of water/GJ of energy in the biomass, while the average water to grow energy crops is approximately 64 m^3 of water/GJ of energy [5]. The data for Table 10.3 were taken from Table 3b in Ref. 5, but Table 10.4 from this same reference lists only 58 m^3 of water/GJ of higher heating value energy as the US energy crop average, so we use an average of these two estimates or 61 m^3/GJ. The

[3]Alkaline electrolyzers require water cooling, but proton exchange membrane (PEM) electrolyzers may not require cooling water.

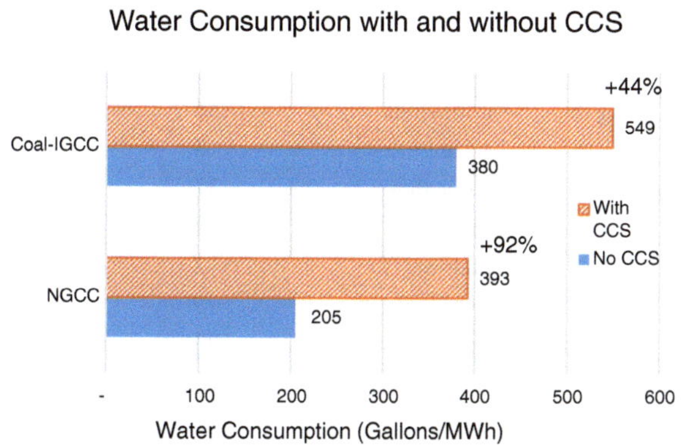

Fig. 10.3 Increased water consumption required for CCS using cooling towers

Table 10.2 Water consumed in making hydrogen in gallons/kg of hydrogen produced

Water req'd for–>	Process	Cooling	Crop water	Totals	Ratios to central SMR
Wind-Alkaline electrolysis-NREL	2.9	294		297	61.2
Wind PEM electrolysis-ANL	7.0	–		7	1.4
Central NG SMR	3.35	1.50		4.9	1.0
Distributed NG SMR	1.52			1.5	0.3
Biomass	1.32		2622	2624	541
IGCC + CCS-NREL	2.91			2.9	0.6
Coal Gasification-ANL	8.50			8.5	1.8

amount of biomass required for hydrogen production will also depend on the efficiency of the gasification plants. Table 10.4 lists the estimated "energy ratios[4]" for four different biomass gasification plants[5] taken from Williams et al. [6]. The average energy ratio of these four plants is 0.737, so the water use must be divided by this factor to determine the amount of biomass and water needed to make a kilogram of hydrogen. The 61 $m^3$of water/GJ (HHV) corresponds to a total water

[4]The "energy ratio" is the ratio of energy in the hydrogen divided by the energy in the biomass input to the plant; since this measure does not include other gasification plant energy inputs such as electricity, it is not a measure of total plant efficiency.

[5]IGT = Institute of Gas Technology; MTCI = Manufacturing and Technology Conversion International; BCL = Battelle Columbus Laboratory.

Table 10.3 Water consumed in growing various crops in the U.S. in cubic meters of water per GJ of biomass energy

	m^3/GJ	
Cotton	135	
Groundnuts	58	
Maize	18	
Miscanthus	37	37
Poplar	42	42
Potatoes	32	
Soybeans	99	
Sugar beats	23	
Sugarcane	30	
Sunflower	61	
Wheat	84	
Oilseedrape	113	113
Averages	61.0	64.0

Table 10.4 Biomass gasifier energy ratios for hydrogen production

IGT	0.669
MTCI	0.759
BCL	0.732
Shell	0.788
Ave.	0.737

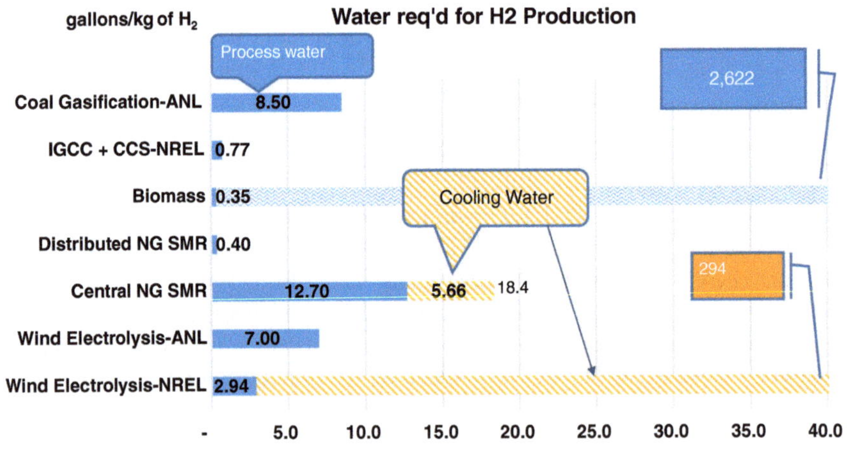

Fig. 10.4 Water required for the production of hydrogen

consumption for crop production of 2624 gallons of water/kg of hydrogen produced with the average 0.737 gasification plant energy ratio.

The final water consumption rates for five different hydrogen production options are shown in Fig. 10.4. The biomass and Wind electrolysis water rates are off this chart to more clearly show the lower water consumption options.

10.4 Water Reduction Potential with Hydrogen and Coal

If society could shift from our current energy sources to a future energy regime based primarily on coal and hydrogen, then water consumption could be reduced substantially

10.4.1 Water Reduction in Electricity Production

As described in Sect. 10.2 above, water consumption from generating electricity could be reduced if society made most electricity from coal at an integrated gasification combined cycle (IGCC) power plant with carbon capture and storage/sequestration (CCS). As shown in Table 10.5, the U.S. would reduce water consumption for electricity generation by 1444 billion gallons per year, a 39 % reduction.

10.4.2 Water Reduction in Transportation Fuel

Significant water is required to produce motor gasoline. The Argonne National Laboratory [7] estimates that between 2.9 and 6.2 gallons of water (average: 4.57 gallons) are required to produce one gallon of gasoline as summarized in Table 10.6

Hydrogen requires less water per unit energy. With our proposed hydrogen production from coal-based IGCC plants with CCS, the average water required is 11.02 L/kg of hydrogen or 2.91 gallons/kg of hydrogen. Given that one gallon of gasoline has about the same energy as one kg of hydrogen,[6] then the water required for gasoline production is approximately 1.57 times larger than for hydrogen production per unit energy.

In addition, the hydrogen-powered FCEV also has much higher fuel economy than a gasoline vehicle, do less energy and hence less water is required for hydrogen per mile traveled. The US Department of Energy National Laboratories

---

[6]One gallon of gasoline has an energy content of 113,602 (LHV) and 121,848 (HHV), compared to one kg of hydrogen at 113,725 (LHV) to 134,510 (HHV).

Table 10.5 U.S. electricity and water consumption for 2013 and estimated savings in water consumption from electricity generation if all electricity came from coal IGCC + CCS

2013 U.S. Electricity and water consumption	Billion kWh	Gallons/MWh	Billion Gallons/yr
Coal	1589.8	1005	1598
Petroleum	17.2		
Natural gas 3/	888.2		
NG ST	383.7	826	317
NG CC	504.5	205	103
Nuclear power	758.1	672	509
Renewable sources			
Hydroelectric	257.0	4491	1154
Biomass	54.3	553	30
Wind	151.6	0	–
Other renewables	14.3	406	6
Total	4141.6		3718
Coal-based IGCC + CCS	4141.6	549	2274
Water savings			*1444*

Table 10.6 Water consumption in the production of gasoline (gallons of water per gallon of gasoline)

	Gallon water/gallon gasoline		
Water use in gasoline production	Min	Max	Ave
U.S. Conventional crude	3.4	6.6	5.0
Saudi conventional crude	2.8	5.8	4.3
Canadian oil sands	2.6	6.2	4.4
Averages	2.93	6.20	4.6

measured the on-road fuel economy of two Toyota FCEVs on their Highlander SUV platform, and determined that the Highlander FCEV has a fuel economy of 68.3 miles/kg of hydrogen [8]

However, this FCEV on a sport utility vehicle (SUV) platform will usually have higher fuel economy on a smaller, lighter mid-size passenger vehicle.

For example, the Highlander has a higher drag coefficient, higher mass and larger cross sectional area than a mid-size car like the Toyota Prius. By taking the fuel cell system out of the Highlander SUV and packaging it into a mid-size car chassis like the Prius would increase the fuel economy from 68.3 mile/kg to as much as approximately 85.4 miles/kg as shown in Table 10.7. We have previously developed a vehicle simulation program under contract to the Ford Motor Company, part of their contract in the mid-1990's to develop their first FCEV. We showed that the EPA driving cycles used to estimate vehicle fuel economy is too anemic, and does not reflect the more aggressive driving habits of American

Table 10.7 Illustration of incremental increases in the Highlander FCEV fuel economy by changing vehicle body parameters to match the body characteristics of the Toyota Prius vehicle

	Cd	Area (ft^2)	Curb wgt (kg)	Miles/kg
Highlander FCHV-adv.	0.326	33	1880	68.3
Prius FCEV reduced mass	0.326	33	1679	71.8
Prius FCEV reduced area	0.326	28	1679	77.2
Prius FCEV reduced Cd	0.25	28	1679	85.4

drivers; we showed that each velocity increment in the EPA urban and highway driving cycles had to be increased by a factor of 1.25 to match real-world driving conditions [9]. We have used this vehicle simulation program to estimate the improvement in fuel economy for reduced mass, reduced cross section and reduced aerodynamic drag. Making the following three changes increases the FCEV fuel economy in steps according to the simulation program:

- Decreasing the vehicle mass from 1880 kg for the Highlander to 1679 for the FCEV on a Prius-like body[7] increases fuel economy to 71.8 miles/kg.
- Reducing the cross sectional body area from 33 ft^3 for the Highlander down to 28 ft^3 for a Prius-like body further increases the fuel economy to 77.2 miles/kg, and, finally
- Reducing the aerodynamic drag coefficient from 0.326 for the Highlander down to 0.25 for a Prius-like body increases the fuel economy to 85.4 miles/kg of hydrogen[8]

Note also that this 85.4 miles/kg fuel economy estimate is a lower bound for several reasons:

1. We have assumed that the fuel cell system for the lighter weight FCEV on a Prius platform weighs the same as the fuel cell system on the Highlander FCEV. With a lighter weight vehicle chassis, the fuel cell power can be scaled down, which would reduce the fuel cell system mass below the 299 kg assumed above; this reduced mass in turn would require less stored hydrogen, further reducing the mass of the hydrogen tank system.
2. We have assumed no increase in the fuel cell system fuel economy, meaning that the fuel cell technology has not improved over the last decade since the Highlander FCEV was built and tested; any improvement in fuel cell system

[7]The Highlander FCEV has a curb mass of 1880 kg, while the conventional gasoline Highlander had a curb mass of 1581 kg. We infer that the difference of 299 kg is due to the fuel cell system (including the fuel cell stack itself, the hydrogen storage system and the peak power battery, net any decreases due to eliminating the internal combustion engine, transmission, etc. Adding this 299 kg net increase for the fuel cell system to a Prius with a curb mass of 1380 kg yields a Prius-body FCEV curb mass of 1679 kg.

[8]However, the current Toyota Mirai FCEV, which is a smaller sedan, has an estimated fuel economy of 66 miles/kg, much lower than the fuel economy of their FCEV on a Highlander platform.

Table 10.8 Water required for conventional gasoline vehicles and fuel cell electric vehicles with hydrogen made from coal IGCC plants with CCS; "Highlander" is the initial Toyota SUV FCEV; "Model" refers to the simulation model developed for the Ford Motor Company with a FCEV on a Prius-like car body

	Gasoline ICVs	
VMT (Million miles/yr)	2,763,000	
Gasoline sold (Quads)	15.307	
Gasoline sold (Million gallons/yr)	134,743	
Water req'd per gallon of gasoline	5.29	
Water consumed (Million gallons/year)	615,324	
H2 FCEVs ==>	Highlander	Model
FCEV fuel economy (miles/kg)	68.3	85.4
H2 req'd per year (Million kg)	40,454	32,354
Water req'd (gallons/kg)	0.769	0.769
Total water consumed for H_2 (Million gallons/year)	31,118	24,887
Ratio of gasoline/hydrogen water req'd	19.77	24.72

efficiency would also increase fuel economy above the 85.4 miles/kg derived above.

3. We have also assumed no increase in the fuel cell system specific power; as with fuel economy, we assume that the fuel cell system mass per unit power will have decreased over the last decade since the Highlander SUV FCEV was built and tested by the National Laboratories.

These three elements will most likely increase the fuel economy above the 85.4 miles/kg shown in Table 10.7.

As shown in Table 10.8, the water required to make gasoline is 20–25 times more than the water consumed making hydrogen from a coal-based IGCC system with CCS for use in a FCEV for the same number of miles traveled.

10.4.3 Summary of Water Savings

From Sect. 4.1, replacing all electricity with coal-based IGCC plus CCS would save 1444 billion gallons of water per year, while replacing all gasoline-powered light duty vehicles with FCEV running on hydrogen from these plants would save 682 billion to 688 billion gallons per year. Thus replacing electricity generators with IGCC plus CCS saves approximately 2.1 times more water than replacing all light duty motor vehicles with hydrogen-powered FCEVs running on hydrogen made at an IGCC coal plant.

References

1. The United Nations World Water Development Report (WWDR) (2014) vol 1 Water and energy. Available at: http://www.unesco.org/new/en/natural-sciences/environment/water/wwap/wwdr/
2. Meldrum J et al (2013) Life cycle water use for electricity generation: a review and harmonization of literature estimates. Environ Res Lett 8 015031. Available at: http://iopscience.iop.org/1748-9326/8/1/015031/article
3. Macknick J, Newmark R, Heath G, Hallett KC (2012) Operational water consumption and withdrawal factors for electricity generating technologies: a review of existing literature. Environ Res Lett 7:045802. Available at: http://iopscience.iop.org/1748-9326/7/4/045802/article
4. Ruth M et al (2009) Hydrogen pathways: cost, well-to-wheels energy use, and emissions for current technology status of seven hydrogen production, deliver and distribution scenarios. Technical report NREL/TP 4A1-46612, National Renewable Energy Laboratory, Sept 2009
5. Gerbens-Leenes P et al (2009) The water footprint of energy from biomass: a quantitative assessment and consequences of an increasing share of bio-energy in energy supply. Ecol Econ 68:1052–1060
6. Williams R et al (1995) Methanol and hydrogen from biomass for transportation, with comparisons to methanol and hydrogen from natural gas and coal. Center for Energy and Environmental Sciences, Princeton University, PU/CESS Report No. 292, July 1995
7. Wu M, Chiu Y (2008) Consumptive water use in the production of ethanol and petroleum gasoline—2011 Update. Argonne National Laboratory, Dec 2008 (updated July 2011) ANL/ESD/09-1-update. Available at: https://greet.es.anl.gov/publication-consumptive-water
8. Wipke K et al (2009) Evaluation of range estimates for the Toyota FCHV-adv. Under open toad driving conditions. The National Renewable Energy Laboratory and the Savannah River National Laboratory, Report # SRNL-STI-2009-00446, 10 Aug 2009. Available at: http://energy.gov/sites/prod/files/2014/03/f9/toyota_fchv-adv_range_verification.pdf
9. Thomas CE (2000) On future fuels: a comparison of options. In: C-J Winter (ed) Chap. 5 of On energies-of-change-the hydrogen solution. Gerling Akademie Verlag, München, Germany

Chapter 11
Coal Mine Safety

Abstract When some of our colleagues learned that we were advocating the use of coal as our primary source of fuel for both electricity and hydrogen for motor vehicles, many were concerned about the danger of coal mining. But we discovered that coal mining safety has improved over the years, and that many other vocations are more dangerous. Furthermore, ending our dependence on imported oil should reap benefits in terms of reduced military actions to safeguard our supply of petroleum in the future and therefore the potential for reduced military injuries and deaths.

11.1 The Dangers of Coal Mining

Coal mining certainly was dangerous with 3292 U.S. coal miners losing their lives in 1907. However, coal mines have become much less dangerous over the years as shown in Fig. 11.1, with only 14 mining deaths recorded in 2014, even as coal production has increased substantially [1]

In addition, coal mining is much less dangerous than nine other occupations as listed in Table 11.1 in terms of deaths per 100,000 people [2], showing that there are nine other occupations with higher death rates than coal mining.[1] Thus logging is almost 8 times more dangerous than coal mining. Even commercial fishing is almost 6 times more dangerous.

[1]Note: the Bureau of Labor Statistics did not even include coal mining on their list of the most "high risk" occupations; we added the Coal miners number on Table 11.1.

C.E. Thomas, *Stopping Climate Change: the Case for Hydrogen and Coal*, Lecture Notes in Energy 35, DOI 10.1007/978-3-319-31655-0_11

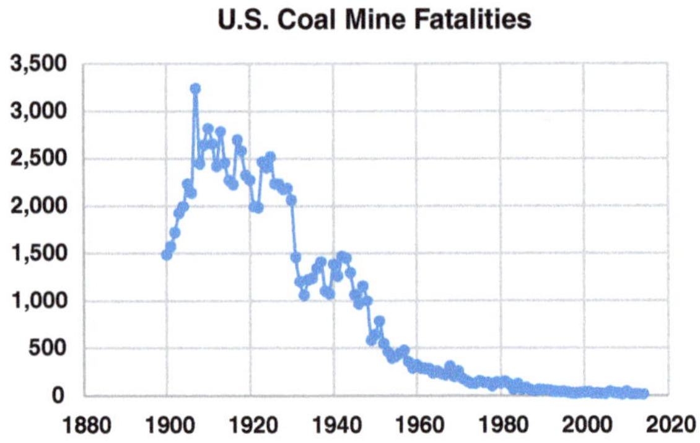

Fig. 11.1 The number of deaths per year in coal mining accidents

Table 11.1 2014 death rates per 100,000 people and ratio of those rates to the coal mining death rate of 13.8 per 100,000 in 2014

	Deaths/100,000	Ratio to miners
Logging	109.5	7.9
Fishermen	80.8	5.9
Pilots	63.2	4.6
Roofers	46.2	3.3
Garbage collectors	35.8	2.6
Farmers, ranchers	26	1.9
Steel workers	25.2	1.8
Truck drivers	23.4	1.7
Electrical power line workers	19.2	1.4
Miners	13.8	1.0

11.2 The Military Dangers of Protecting Our Sources of Petroleum

In addition, moving toward an economy fueled primarily by domestic coal would improve our national security by eliminating the need to protect our supplies of imported oil. This should reduce deaths from military operations in the Middle East.

The death rates[2] in U.S. military operations in Iraq and Afghanistan are shown in Table 11.2 [3] These military death rates after 2002 are also higher than the coal mine rate of 13.8 deaths per 100,000 miners in 2014.

[2]These military death rates are the deaths divided by the *entire* U.S. military, not just the troops serving in the war zone. So the death rates for soldiers in the combat zones are much higher.

Table 11.2 U.S. military deaths in Iraq and Afghanistan and death rates per 100,000 troops in the total U.S. military (not just the military in the war zone)

	Afghan	Iraq	Totals	Deaths/100,000
2002	49		49	3.5
2003	48	486	534	37.5
2004	52	849	901	63.8
2005	99	846	945	68.6
2006	98	823	921	66.7
2007	117	904	1021	74.0
2008	155	314	469	31.9
2009	317	149	466	31.7
2010	499	60	559	38.1

In summary, implementing this plan to produce most electricity and most motor fuel (as hydrogen) from coal should reduce military deaths more than it increases coal mining deaths.

11.3 Black Lung Disease

Unfortunately, mining accidents are not the major cause of death in the Coal Industry. In 2004, for example, 25 times more miners died of Black Lung disease (703) than died in mining accidents (28) [4]. Fortunately the concentration of coal dust and Silica has been declining with increased regulation and monitoring of coal dust in the mines. According the U.S. Department of Labor Mine Safety and Health Administration, dust concentration in mines fell from 0.82 mg/m^3 in 2009 to 0.68 mg/m^3 in 2015 [5]. Hopefully this trend will continue to reduce the affects of Black Lung Disease.

References

1. Coal Fatalities for 1900 through 2014, by the U.S. Department of Labor. Available at: http://arlweb.msha.gov/stats/centurystats/coalstats.asp
2. Civilian occupations with high fatal work injury rates-2014. U.S. Bureau of Labor Statistics. Available at: http://www.bls.gov/iif/oshwc/cfoi/cfch0013.pdf
3. Number of coalition fatalities among Western coalition soldiers involved in the execution of Operation Enduring Freedom from 2001 to 2015. Available at: http://www.statista.com/statistics/262894/western-coalition-soldiers-killed-in-afghanistan/
4. NPR; the quiet deaths outside the coal mines. Available at: http://www.npr.org/2010/04/16/126021059/the-quiet-deaths-outside-the-coal-mines
5. Underground coal mines dust concentration by calendar year. U.S. Mine Safety and Health Administration. Available at: http://arlweb.msha.gov/endblacklung/dust-concentration-by-year-01-19-2016.pdf

Chapter 12
Conclusions: "Stopping Climate Change: The Case for Coal and Hydrogen"

Abstract This book demonstrates that by shifting electricity production from coal, natural gas and oil to coal gasification (with carbon capture and sequestration), nuclear power, and renewables, and shifting from gasoline to hydrogen made from coal to power most vehicles, we can eliminate most local air pollution, meet or exceed societal goals for reducing greenhouse gas emissions by at least 80 % below 1990 levels by 2050, reduce our consumption of water, and virtually eliminate our dependence on imported oil, thereby significantly improving national security.

The main conclusions from this book may be quite surprising to much of the public:

1. The consequences of climate change are far more threatening than most citizens realize: while many people associate climate change with rising sea levels and more severe weather events including flooding, droughts and firestorms, the threat of rising global population growth combined with reduced farmland suitable for growing food crops could severely hamper our ability to feed the expanding global population and result in the exodus of many tens of millions of people unable to feed their families from Africa, Asia and South America to Europe and the United States, which may make the refugee crisis of 2014–2015 due to Middle East violence seem tame by comparison.
2. Coal, used appropriately, is a potential *antidote* to climate change, not the main *antagonist*.
3. The advantages of coal used in an integrated gasification combined cycle (IGCC) electricity generator with carbon capture and storage (CCS) include:
    a. Sharply reduced greenhouse gases (87.4 g/kWh) compared to current coal-powered steam turbines (11115 g/kWh) or even natural gas combined cycle (NGCC) power plants (507 g/kWh).
    b. The ability to produce hydrogen from coal to run most light-duty vehicles, heavy-duty trucks, trains and ships using fuel cells.

C.E. Thomas, *Stopping Climate Change: the Case for Hydrogen and Coal*,
Lecture Notes in Energy 35, DOI 10.1007/978-3-319-31655-0_12

c. The possible elimination of *all* imported oil[1] as most vehicles run on hydrogen instead of oil products, dramatically improving national security for most nations of the world (including U.S.) that have ample coal reserves but very limited oil resources.
d. Hydrogen made from coal and used in fuel cell electric vehicles (FCEVs) in turn has these advantages over gasoline or diesel as a motor fuel:

   i. Dramatically reduced greenhouse gas emissions compared to gasoline or diesel powered vehicles, battery-powered electric vehicles,[2] or hybrid electric vehicles.
   ii. The ability to store intermittent renewable energy as hydrogen for later use when the sun is not shining and the wind is not blowing, thereby enabling greater utilization of intermittent renewable energy.
   iii. The ability to earn a 30-year return on investment of over 12 % in Los Angeles and over 15 % in Hawaii as described in Chap. 8.

e. Electricity and hydrogen made from a coal-powered IGCC with CCS also consumes less water than current fuel processing systems such as power plants and oil refineries, thereby reducing the stress on the world's limited freshwater supplies.

In Summary, if we gradually build coal-powered integrated combined cycle power (IGCC) plants with carbon capture and storage (CCS) over the next few decades to provide both electricity and hydrogen to fuel most vehicles, then we can:

- Meet or exceed the climate change goal of cutting greenhouse gases by 80 % below 1990 levels by 2050
- Eliminate all imported oil
- Substantially reduce freshwater consumption for electricity and gasoline production.
- Install more intermittent renewables backed up by stored hydrogen when the sun is not shining and the wind is not blowing.

In addition to the main long-term conclusions described above, we also identified one shorter term project to profitably jump-start the hydrogen infrastructure to begin converting from gasoline to hydrogen-powered vehicles, and one longer-term project to profitably combine solar energy with hydrogen storage.

As described in Chap. 7, society can begin profitably installing hydrogen fueling systems at "Big Box" stores and warehouses that have already installed hydrogen-powered fuel cell Material Handling Equipment (MHE) such as forklift trucks. These early hydrogen stations would convert natural gas to hydrogen that would be used to generate electricity via a stationary proton exchange membrane (PEM) fuel cell system, displacing electricity previously purchased from the grid.

[1]The U.S. can produce enough oil domestically to supply all non-transportation uses of petroleum.
[2]With electricity produced predominantly by burning coal.

Some hydrogen would be used to power fuel cell MHEs and FCEVs. We showed that as long as the Big Box store or warehouse had more than 100 fuel cell MHEs, the 20-year return on investment would be over 10 %. The 36 warehouses analyzed here had an average of 160 fuel cell MHEs, and would earn more than 15 % return on investment over 20 years.

For the longer term, hydrogen could be generated from photovoltaic (PV) solar panels and an electrolyzer. We analyzed two photovoltaic (PV) systems to provide zero greenhouse gas electricity in Los Angeles in Chap. 9: a fixed tilt PV system and a single-axis tracking system, both generating and storing hydrogen to provide electricity when the sun is not shining. We also analyzed two other options to back up intermittent solar energy: natural gas combined cycle (NGCC) plants and nuclear power plants. NGCC plants are the least costly option, but these natural gas-powered plants used to backup solar plant would generate *four times more* greenhouse gases than needed to meet our GHG goal of cutting GHGs by 80 % below 1990 levels by 2050, even if PV-solar energy replaced all fossil fuel power plants. Nuclear power plants would allow us to meet the GHG goal, but they would carry the burden of public acceptance, nuclear waste disposal and risk of nuclear weapons proliferation.

These short- and long-term projects of the hydrogen-energy-from-coal plan presented in this book are at once practical yet sweeping in ramifications to once and for all meet our society's energy, climate, environmental, and energy-related national security goals.

Appendix A
The Economics of IGCC plants with Carbon Capture and Storage

A key premise of this book is that the electric utility industry can afford to replace their existing generation plants with coal-based integrated combined cycle (IGCC) plants with CCS.[1] This appendix explores the economics of building and operating such plants, including the impact of a carbon tax; we show that 20-year IRRs above 11 % are possible in many parts of the US, with a US average above 13 % with a carbon fee/tax.

1. *Electricity generation plant costs*

The key parameters including the capital costs ($/kWe) of the various combined cycle plants from Chap. 4 are summarized in Table A.1, based on the averages from five literature sources.[2] All capital cost estimates have been escalated to 2015$ using the Power Capital Cost Index (PCCI) [1].

2. *CO_2 transportation and storage costs*
3. The capital costs in Table A.1 for plants with CCS include the cost of extracting and pressurizing CO_2. But we must also add the costs of transporting the CO_2 to a disposal well, and also the costs of building that well. We use the average costs from five literature sources as summarized in Table A.2.
4. *Plant operating costs*

    a. *IGCC coal purchase costs*

The annual cost of coal for each plant will depend in the net efficiency and the capacity factor of the plants. From Table A.1, the average estimated HHV efficiency of an IGCC+CCS plant is 34.7 %. We assume an average annual capacity factor of 95 % for these baseload power plants. In this model, we use the estimated delivered cost of coal from the Energy Information Administrations' (EIA) 2015 Annual Energy Outlook (AEO) [7]. Average[3] U.S. Coal costs increase from an

[1]We assume that *all* IGCC plants must have CCS to meet our GHG reduction goal.
[2]See Chap. 4 for references.
[3]We use the average US coal prices only for the US case, but use the projected coal prices for the other regions of the country.

C.E. Thomas, *Stopping Climate Change: the Case for Hydrogen and Coal*, Lecture Notes in Energy 35, DOI 10.1007/978-3-319-31655-0

Table A.1 Average characteristics of various combined cycle plants from Chap. 4

	IGCC	IGCC+CCS	NGCC	NGCC+CCS
Capex ($/kWe)	$2866	$4207	$1003	$1811
GHG (g/kWh)	1112.1	87.4	507.2	50.9
HHV efficiency (%)	40.8	34.7	51.9	45.4

Table A.2 Cost of transportation (100 miles) and storage of CO_2 in $/metric tonne

$/tonne CO_2	Transport	Storage
Ogden [2]	5.55	1.78
CMU [3]	1.16	0.95
NETL [4]	3.65	11.81
PNNL [5]	3.52	11.48
MIT [6]	0.62	9.38
Average	2.90	7.08

estimate of $2.25/MBTU in 2016 up to $2.92/MBTU by 2040 in the AEO-2015 forecast.

b. *IGCC operations and maintenance costs*

The fixed and variable operation and maintenance (O&M) costs are taken from an average of three estimates in the literature as summarized in Table A.2. "NREL-'10 refers to a 2010 report by Tidball et al. [8], AEO'15 is from Table 8.2 of the AEO-2015 [7], and EIA'13 is from a report by Black and Veatch for the National Renewable Energy Laboratory (NREL) [9].

5. *Electricity generation plant revenues*

We use the average electricity price from the EIA's 2015 Annual Energy Outlook in this model to calculate the revenue for the utility building this 200-MW plant, assuming an average capacity factor of 95 %. THE EIA estimates an average electricity price (averaged over all sectors) of 10.9 cents/kWh in 2016, increasing to 19.2 cents/kWh by 2040. But these prices include the cost of transmission and distribution (T&D) which will not accrue to the owner of the IGCC+CCS plant. We must therefore subtract the T&D costs from the average electricity price. The average U.S. T&D costs varied from 4.04 ₵/kWh in 2016 to 6.97 ₵/kWh in 2040, so the net electricity price for the generator varied from 6.67 ₵/kWh in 2016 to 12.21 ₵/kWh in 2040.

6. *Carbon taxes*

We assume that carbon taxes will eventually be enacted to help curb the threats of climate change. Carbon taxes will benefit the IGCC+CCS plant relative to the existing Coal steam turbine and NGCC plants, since IGCC+CCS plants have lower GHG emissions. The price of electricity is determined primarily in competitive wholesale markets in the U.S. Each generator bids into a day-ahead market, with the generators with the lowest operating costs (including carbon taxes) setting the "clearing price" for the next day period; the lowest cost generator that just meets the load for the period on the next day sets the "clearing price," which is then the price paid to *all* generators. Carbon taxes, although paid by the consumer, are included in the bid price. So the owner of the generator plant receives revenues equal to the bid price minus the carbon taxes. As a result, low GHG plants like the IGCC+CCS plants with 87.4 g/kWh GHGs will receive slightly more revenue than either Coal steam turbine plants (1115/kWh) or NGCC plants (507 g/kWh).

Figure A.1 shows the six carbon tax plans introduced in Congress plus one plan proposed by the Citizens' Climate Lobby (CCL). The CCL plan is a revenue-neutral plan based on a carbon "fee" that is fully refunded to citizens in the form of rebates [10]. We use the CCL plan here as representative of the upper bound on Congressional plans introduced to date. The impact of the CCL plan is illustrated in Table A.3 for the first year, when the carbon fee would be only \$16.54 per tonne of CO_2-equivalent. The IGCC+CCS plant would only have an advantage of 1.2 ₵/kWh over the average of the Coal-ST and NGCC plants. This may not seem significant, but it grows over the years as the carbon fees increase, and this small difference increases the return on investment in the IGCC+CCS plant as shown below.

7. *IGCC cash flow calculations*

The key assumptions for a 200-MW IGCC+CCS plant are summarized in Table A.4.

The rate of return on the investment depends on the average prices of electricity and coal that vary widely across the U.S. as shown in Table A.5. Regions with low

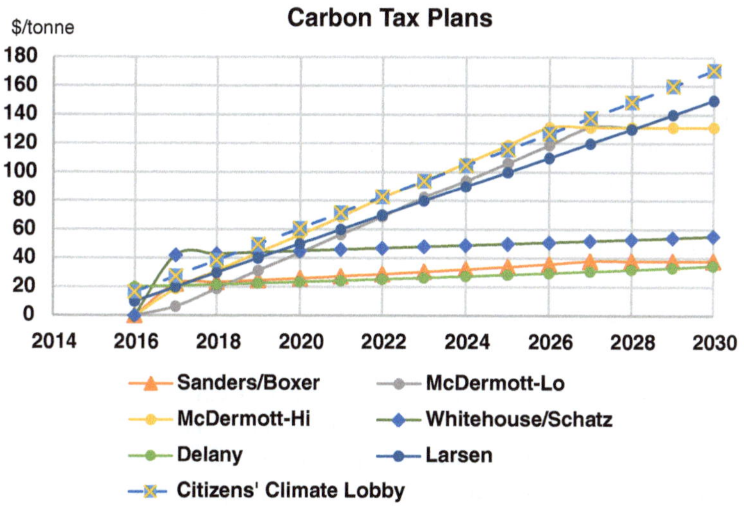

Fig. A.1 Six carbon tax plans introduced in congress plus one citizens' plan

Table A.3 Estimates of the average fixed and non-fuel variable operation and maintenance (O&M) costs from three literature sources

	NREL-'10	AEO '15	EIA '13	Average:
FIXED O&M ($/Kw/YR)				
Coal-ST	27.53	31.16	37.8	32.16
IGCC	38.67	$51.37	62.25	50.76
IGCC+CCS	46.12	$72.80	72.83	63.92
NGCC	12.48	$15.36	15.37	14.40
NGCC+CCS	19.9	$31.77	31.79	27.82
Nuclear	0.49	93.23	93.28	62.33
Variable O&M ($/MWh)				
Coal-ST	4.59	4.47	4.47	4.51
IGCC	2.92	$7.22	7.22	5.79
IGCC+CCS	4.44	$8.44	8.45	7.11
NGCC	2.07	$3.27	3.27	2.87
NGCC+CCS	2.94	$6.78	6.78	5.50
Nuclear	90.02	$2.14	2.14	31.43

Table A.4 Comparisons of the net revenue of the IGCC+CCS plant compared to the existing Coal-ST and NGCC plants

	Clearing price (₵/kWh)	GHGs (g/kWh)	Carbon tax		Net generator "Revenue" (₵/kWh)
			($/tonne)	(₵/kWh) (c/kWh)	
Coal-ST	10.00	1115.3	$16.54/tonne	1.84	8.16
NGCC	10.00	507.2	$16.54/tonne	0.84	9.16
IGCC +CCS	10.00	87.4	$16.54/tonne	0.14	9.86

Table A.5 Key assumptions for a 200-MW IGCC+CCS power plant cash flow calculations

Plant size (MW)	200
Plant cost ($/kW)	$4207/kW
Plant capacity factor	95 %
Fixed O&M ($/kW-year)	63.92
Variable O&M ($/MWh)	7.11
Plant efficiency	34.7 %
Construction finance charge	12 %

coal costs and high electricity costs will favor IGCC+CCS plants. Hawaii is the best location financially for an IGCC plant, since Hawaii has the highest electricity prices in the nation and also surprisingly low coal prices, as shown in Table A.6 for 2015. Hawaii does not have any coal reserves, but imports coal from Indonesia to operate one 180-MW coal plant on Oahu at a cost of only $3.94/MBTU in 2015,

Table A.6 Electricity and coal prices by region and ratio of generator electricity price to coal price

2015	Total electricity (₵/kWh)	Electricity T&D (₵/kWh)	Generator electricity (₵/kWh)	Generator electricity ($/MBTU)	Coal ($/MBTU)	Ratio (elect/coal)
US average	10.10	3.67	6.43	18.85	2.34	8.04
New England	14.30	7.69	6.61	19.38	3.92	4.95
WECC-rockies	9.18	3.20	5.98	17.54	1.87	9.38
WECC-California	14.27	6.93	7.34	21.52	1.98	10.86
MRC-West	8.40	1.11	7.29	21.38	1.66	12.89
Hawaii	32.98	3.67	29.31	85.90	3.94	21.81

Table A.7 IGCC+CCS plant Internal rates of return for US regions assuming adoption of the citizens' Climate Lobby carbon fee and rebate plan (the Hawaiian columns with multiplier factors indicate that the Hawaiian electricity prices paid to the IGCC plant were reduced by those factors.)

	US average (%)	WECC-California (%)	New England (%)	WECC-rockies (%)	MRC-West (%)	HI x.8 (%)	HI x.5 (%)	HI x.3 (%)
15-year IRR	10.95	13.92	9.18	11.00	14.20	50.7	21.6	6.22
20-year IRR	13.09	15.73	11.61	13.21	15.98	50.8	22.6	8.99
30-year IRR	15.05	17.29	13.91	15.21	17.49	50.8	23.3	11.75

WECC Western Electricity coordinating council
MRC Midwest reliability council

The average US ratio of generator electricity price[4] to coal price is 8.04 to one, but Hawaii's ratio is 2.7 times larger at 21.81 to one.

The resulting rates of return for the owner of the IGCC+CCS plant are shown in Table A.7 for five regions plus the US average. Even the 15-year IRR exceeds 10 % for the U.S. average electricity price projection, with an average 20-year IRR of more than 13 %, which is above the average allowed rate of 10.23 % from nine rate cases decided in the first quarter of 2014 [11]. In other words, the utility as a regulated "natural monopoly" would not be allowed to charge their customers the electricity rates that resulted in the higher returns listed in Table A.6. However, most U.S. electricity today is sold in competitive wholesale markets that do not have a cap on allowed rate of return. In the case of Hawaii, the price of electricity

[4]The generator electricity price is the all-sector weighted average price minus the transmission and distribution (T&D) cost that does not accrue to the generator plant owner.

Table A.8 Detailed cash flow calculation (first 15 years) for a 200-MW IGCC+CCS power plant with projected average US electricity rates and the Citizens' climate lobby carbon fee

	2016	2017	2018	2019	2020	2021	2022
All-sector elect. cost (₵/kWh)- US	10.296	10.338	10.255	10.332	10.512	10.682	10.750
Total operating costs + C. tax	$2,406,300	$4,010,474	$5,614,633	$7,218,792	$8,822,952	$10,427,111	$12,031,270
IGCC+CCS Operating + C. tax cost (₵/kWh)	0.14	0.241	0.34	0.43	0.53	0.63	0.72
NGCC C tax (₵/kWh)	0.83872	1.39786	1.95700	2.51615	3.07529	3.63444	4.19358
Natural gas costs-utility ($/MBTU)	4.42	4.40	4.53	4.96	5.39	5.65	5.80
Fraction of year operating time	1.00	1.00	1.00	1.00	1.00	1.00	1.00
T&D cost (₵/kWh)	3.83	3.90	3.91	3.92	3.95	3.98	4.00
Delta tax advantage	1.20	1.99	2.79	3.59	4.39	5.19	5.98
Price to generator (₵/kWh)	7.66	8.43	9.14	10.00	10.95	11.89	12.73
Coal-utility costs ($/MBTU)	2.3734	2.4413	2.5197	2.6092	2.7005	2.7862	2.8736
CO_2-eq. emissions (tonnes/year)	145,501.85	145,501.85	145,501.85	145,501.85	145,501.85	145,501.85	145,501.85
Carbon tax ($/tonne) citizens' climate lobby tax	16.54	27.56	38.59	49.61	60.64	71.66	82.69
Carbon tax for coal-ST (₵/kWh)	1.84	3.07	4.30	5.53	6.76	7.99	9.22
Carbon tax for NGCC (₵/kWh)	0.84	1.40	1.96	2.52	3.08	3.63	4.19
Carbon tax (₵/kWh) for IGCC+CCS	0.145	0.24095	0.33733	0.43371	0.53009	0.62647	0.72285
Expenses	200 MW						
Capital cost ($M)	$841.5						
CO_2 transport (100 miles) ($M)	$4.3						
CO_2 injection ($M)	$10.55						
Construction financing (12 %) ($M)	$100.98						
Total ($M)	$957.3						
T&I ($M)	$15	$14.73	$15	$15	$14.7	$15	$15

(continued)

Table A.8 (continued)

	2016	2017	2018	2019	2020	2021	2022
Fixed O&M	$12.78	$12.78	$12.78	$12.78	$12.78	$12.78	$12.78
Variable O&M	$11.83	$11.83	$11.83	$11.83	$11.83	$11.83	$11.83
Coal costs	$38.845	$39.96	$41.24	$42.70	$44.1990	$45.60	$47.03
Total costs ($M)	$1035.5	$79.30	$80.58	$82.05	$83.542	$84.94	$86.37
Annual kWh	1,664,400,000	1,664,400,000	1,664,400,000	1,664,400,000	1,664,400,000	1,664,400,000	1,664,400,000
Fuel MBTU req'd	16,366,822	16,366,822	16,366,822	16,366,822	16,366,822	16,366,822	16,366,822
Revenue-electricity sales (all sectors) ($M)	$127.58	$140.31	$152.14	$166.44	$182.29	$197.83	$211.94
Cash flow ($M))	$(907.9)	$61.0	$71.6	$84.4	98.8	$112.9	$125.6
15-year IRR	10.95 %						
20-year IRR	13.09 %	US					
30-year IRR	15.05 %	IGCC+CCS	Citizens' climate lobby tax/fee				

	2023	2024	2025	2026	2027	2028	2029	2030
All-sector elect. cost (₵/kWh)- US	10.833	10.842	10.955	10.999	11.049	11.075	11.044	11.050
Total operating costs + C. tax	$13,635,430	$15,239,589	$16,843,748	$18,447,907	$20,052,067	$21,656,226	$23,260,386	$24,864,545
IGCC+CCS Operating + C. tax cost (₵/kWh)	0.82	0.92	1.01	1.11	1.20	1.30	1.40	1.49
NGCC C tax (₵/kWh)	4.75272	5.31187	5.87101	6.43015	6.98930	7.54844	8.10759	8.66673
Natural gas costs-utility ($/MBTU)	5.97	6.10	6.27	6.44	6.40	6.28	6.26	6.22
Fraction of year operating time	1.00	1.00	1.00	1.00	1.00	1.00	1.00	1.00
T&D cost (₵/kWh)	4.01	4.02	4.05	4.07	4.08	4.10	4.12	4.14
Delta tax advantage	6.78	7.58	8.38	9.18	9.97	10.77	11.57	12.37
Price to generator (₵/kWh)	13.61	14.40	15.29	16.11	16.94	17.75	18.49	19.28

(continued)

Table A.8 (continued)

	2023	2024	2025	2026	2027	2028	2029	2030
Coal-utility costs ($/MBTU)	2.9612	3.0416	3.1274	3.2124	3.3041	3.3953	3.4870	3.5794
CO_2-eq. emissions (tonnes/year)	145,501.85	145,501.85	145,501.85	145,501.85	145,501.85	145,501.85	145,501.85	145,501.85
Carbon tax ($/tonne) citizens' climate lobby tax	93.71	104.74	115.76	126.79	137.81	148.84	159.86	170.89
Carbon tax for coal-ST (₵/kWh)	10.45	11.68	12.91	14.14	15.37	16.60	17.83	19.06
Carbon tax for NGCC (₵/kWh)	4.75	5.31	5.87	6.43	6.99	7.55	8.11	8.67
Carbon tax (₵/kWh) for IGCC+CCS	0.81923	0.91562	1.01200	1.10838	1.20476	1.30114	1.39752	1.49390
Expenses								
Capital cost ($M)								
CO_2 transport (100 miles) ($M)								
CO_2 injection ($M)								
Construction financing (12 %) ($M)								
Total ($M)								
T&I ($M)	$15	$15	$15	$15	$15	$15	$15	$15
Fixed O&M	$12.78	$12.78	$12.78	$12.78	$12.78	$12.78	$12.78	$12.78
Variable O&M	$11.83	$11.83	$11.83	$11.83	$11.83	$11.83	$11.83	$11.83
Coal costs	$48.46	$49.78	$51.19	$52.58	$54.08	$55.57	$57.07	$58.58
Total costs ($M)	$87.81	$89.12	$90.53	$91.92	$93.42	$94.91	$96.41	$97.93
Annual kWh	1,664,400,000	1,664,400,000	1,664,400,000	1,664,400,000	1,664,400,000	1,664,400,000	1,664,400,000	1,664,400,000

(continued)

Table A.8 (continued)

	2023	2024	2025	2026	2027	2028	2029	2030
Fuel MBTU req'd	16,366,822	16,366,822	16,366,822	16,366,822	16,366,822	16,366,822	16,366,822	16,366,822
Revenue-electricity sales (all sectors) ($M)	$226.51	$239.66	$254.45	$268.14	$281.93	$295.35	$307.83	$320.87
Cash flow ($M))	$138.7	$50.5	$163.9	$76.2	$188.5	$200.4	$211.4	$222.9
15-year IRR								
20-year IRR								
30-year IRR								

could be reduced by up to 70 % (shown by "HI x .3" in Table A.7) and the IGCC +CCS plant would still make a 30-year IRR above 11 % in Hawaii.

Finally, the details of the IGCC+CCS plant cash flows are shown in Table A.8 on the next page for the US average electricity prices.

References

1. The IHS Power Capital Cost Index (PCCI) for North America. Available at https://www.ihs.com/info/cera/ihsindexes/
2. Ogden J (2002) Modeling infrastructure for a fossil hydrogen energy system with CO_2 sequestration. In: Sixth greenhouse gas control technologies conference, Kyoto, Japan, 30 Sept–4 Oct 2002
3. Rubin E et al (2008) Technical documentation: the economics of CO_2 transport by pipeline storage in saline aquifers and oil reserves. Department of Engineering and Public Policy, Carnegie Mellon University
4. Grant T et al (2013) Carbon dioxide transport and storage costs in NETL studies. DOE/NETL-2013/1614, Final report, 14 Mar 2013
5. Dooley J et al (2008) On the long-term average cost of CO_2 transport and storage. Pacific Northwest National Laboratory, PNNL-17389
6. Herzog H (2010) Scaling up carbon dioxide capture and storage: from megatons to gigatons. Energy Economics
7. The Annual Energy Outlook (AEO) for 2015, the U.S. Energy Information Administration. Available at http://www.eia.gov/oiaf/aeo/tablebrowser/
8. Tidball R et al (2010) Cost and performance assumptions for modeling electricity generation technologies. In: ICF International for the national renewable energy laboratory, subcontract report NREL/SR 6A20-48595
9. Cost and performance data for power generation technologies. Black and Veatch for the National Renewable Energy Laboratory
10. Citizens' Climate Lobby "Carbon fee and dividend. Available at https://citizensclimatelobby.org/carbon-fee-and-dividend/
11. Edison Electric Institute, rate case summary, quarterly report of the U.S. shareholder-owned utility industry. Available at http://www.eei.org/resourcesandmedia/industrydataanalysis/industryfinancialanalysis/QtrlyFinancialUpdates/Documents/QFU_Rate_Case/2014_Q1_Rate_Case.pdf

Index

C.E. Thomas, *Stopping Climate Change: the Case for Hydrogen and Coal*,
Lecture Notes in Energy 35, DOI 10.1007/978-3-319-31655-0

MIX
Papier aus verantwortungsvollen Quellen
Paper from responsible sources
FSC® C105338

If you have any concerns about our products,
you can contact us on
ProductSafety@springernature.com

In case Publisher is established outside the EU,
the EU authorized representative is:
Springer Nature Customer Service Center GmbH
Europaplatz 3, 69115 Heidelberg, Germany

Printed by Libri Plureos GmbH
in Hamburg, Germany